Data Analytics and Business Intelligence
Computational Frameworks, Practices, and Applications

Editors:

Vincent Charles
Programme Director for MSc Applied Artificial Intelligence and
Data Analytics School of Management, University of Bradford
Bradford, United Kingdom

Pratibha Garg
Assistant Professor, Amity School of Business
Amity University, Noida, Uttar Pradesh, India

Neha Gupta
Assistant Professor, Amity School of Business
Amity University, Noida, Uttar Pradesh, India

Mohini Agarwal
Unaffiliated Independent Researcher
Glendale, Arizona 85308 United States

CRC Press
Taylor & Francis Group
Boca Raton London New York

CRC Press is an imprint of the
Taylor & Francis Group, an **informa** business

A SCIENCE PUBLISHERS BOOK

First edition published 2023
by CRC Press
6000 Broken Sound Parkway NW, Suite 300, Boca Raton, FL 33487-2742

and by CRC Press
4 Park Square, Milton Park, Abingdon, Oxon, OX14 4RN

Library of Congress Cataloging-in-Publication Data (applied for)

ISBN: 978-1-032-03904-6 (hbk)
ISBN: 978-1-032-03906-0 (pbk)
ISBN: 978-1-003-18964-0 (ebk)

DOI: 10.1201/9781003189640

Typeset in Palatino
by Radiant Productions

Preface

Data Analytics is an evolving phenomenon that showcases the increasing importance of using a huge volume of data and showing how it can be used to generate value for the business. The advances in Data Analytics (BA) have offered great opportunities for organizations to innovate and develop new or improve existing products/services. This book will uncover stimulus for innovative business solutions covering all aspects of business analytics such as management science, information technology, descriptive, prescriptive, and predictive models, machine learning, network science, and mathematical and statistical techniques. The book will present a valuable collection of chapters exploring and discussing the framework, practices, and applications of data analytics and business intelligence that can assist industries in decision-making, problem-solving, and gaining a competitive advantage.

The book has been divided into three sections, wherein the first section is dedicated to chapters based on operations and supply chain analytics and the second section is a compilation of chapters related to data mining, computational framework, and practices, and the third section is based on business intelligence and analytics applications. In the first Chapter, "Wholesale Price Strategy of a Manufacturer under Collusion of Downstream Channel Members: A Game-Theoretic Approach", the wholesale price strategies of the manufacturer have been compared by considering a dual-channel supply chain comprising a downstream Brick and Mortar (BM) store, and an E-store engaged in collusion. By assuming the manufacturer as the channel leader, Stackelberg Game models were designed, and equilibrium analysis was carried out to obtain the profit-maximizing price and corresponding profit of the supply chain members. With the help of a numerical example, it has been highlighted that it is optimal for the manufacturer to supply the product at the same wholesale price to the BM-store and E-store for products of low product-web fit and high product-web fit. Moreover, it is optimal for the manufacturer to supply the product at different wholesale prices to the BM-store and E-store for products of moderate product-web fit.

In Chapter 2, "Artificial Intelligence (AI) and Machine Learning (ML) in Supply Chain Decision Making – A pragmatic Discussion", AI and ML are essential strategic tools used in the predictive analytics domain of supply chain management. The chapter gives an overview of the present trend of using data analytics through AI and ML in making supply chain decisions and its contribution to making the current supply chain lean and agile. The adoption of AI and ML is quite slow but can be quite promising for business growth, provided it is being adopted by making the current system and environment fit for its implementation. This chapter aims to present a theoretical idea of supply chain segmentation based on decision-making by considering dynamic demand factors applicable across various industries.

In Chapter 3, "Assessing Relations of Lean Manufacturing, Industry 4.0 and Sustainability in manufacturing Environment", reflects on how sustainability in the manufacturing environment relates to the production philosophy of Lean Manufacturing (LM) and Industry 4.0. Many organizations are moving towards the adoption of sustainable practices of manufacturing which can be measured by the three pillars economic, social, and environmental. The authors have examined the relationship between production philosophies and sustainability indicators. To evaluate this relationship, a general framework with six hypotheses to measure the effect of these production philosophies on sustainability has been considered. To statistically analyse the hypothesis a set of responses has been collected and it was observed that LM is correlated with all three sustainability pillars, whereas Industry 4.0 is correlated with only two pillars. The result of this study can contribute to an important decision for Indian manufacturing organizations toward the adaptation of LM and Industry 4.0.

In Chapter 4, "Role of Artificial Intelligence in Supply Chain Management", Artificial intelligence (AI) continues to dominate the business and non-business environments amid various criticisms due to the fear that AI technology will endanger the role of people in future management and business operations. In this chapter, the authors seek to identify different approaches through which AI and some other technologies transform the jewellery business. A well-functioning supply chain is a key to success for every business entity. Having an accurate projection of inventory offers a substantial competitive advantage

The fifth chapter, "Impact of Blockchain in Creating Sustainable Supply Chain", highlights how blockchain technology can be a game-changer in building sustainable supply chains with an overall low carbon footprint per unit of product, eliminating possibilities of fraud, double-counting in the carbon credit market and in bringing authentication and transparency in audits and certification of sustainability practices. Blockchain technology will help in building a sustainable supply chain. It can address the challenges of standardization in the measurement of carbon emission across supply chains which would help in reducing the overall carbon footprint across the supply chains. The blockchain can be extremely useful in the carbon credit market by eliminating double counting and multiple claims for the same offset.

The sixth chapter, "Exploring Adoption of Blockchain Technology for Sustainable Supply Chain Management", highlights how the Shared ledger, Smart contract, Privacy, Trust, and Transparency, the five pillars of Blockchain, can ease some of the global difficulties faced in supply chain management and create sustainable supply chains. It further discusses the application of blockchain and challenges in the adoption of blockchain technology.

The second section of this book is based on chapters about Data Mining, Computational Framework, and Practices. The seventh chapter, "Mathematical Model of Consensus and its Adaptation to Achievement Consensus in Social Groups", discusses how this approach ensures that all the opinions of the group members, their ideas, and needs will be considered. This chapter discusses the results of statistical modeling describing the dependence of the time to reach consensus on the number and authoritarianism of social group members using a mathematical model of consensus achievement in a group based on the model proposed by DeGroot. Also, the mathematical model of consensus is constructed on the work of social groups in terms of coalitions, which are overcome during the negotiation process by concessions. It was revealed that even a small concession leads to the onset of consensus, increasing the size of the assignment results in a rapid decline in time before the consensus, which is crucial for practice.

Chapter 8, "Data to Data Science: A Phenomenal Journey", helps the readers to gain insights about the data warehouse, data mining, and big data and its challenges. Data has been crucial for humans since the pre-historic age, from counting to bartering and other activities. With evolution, data has occupied a very important position in human life which can be used for future decision-making. The approach to recording and storing data has moved from the conventional approach to a recent manner of digitalization of data. The digitalization of human society has started generating data at very large volumes, velocities, and variety. This data is of semi-structured form and needs a robust and reliable approach to extract meaningful information. Thus, big data and data science have taken a central role in capturing, analysing, and extracting meaningful information. This chapter describes different techniques and the challenges that can be used for extracting meaningful information.

Chapter 9, "Application of Algorithm on Computational Intelligence and Machine Learning for Product Design: Emerging Needs and Challenges", Computational intelligence is one of the most fascinating techniques that has recently joined the material science toolkit for machining learning. This set of statistical tools has already demonstrated its ability to significantly accelerate both fundamental and practical research. Right now, several attempts are being placed into growing and device studying (machine learning) implemented to solid-nation devices. Machine learning concepts, algorithms, descriptors, and databases are the starting points in materials science. Various machine learning algorithms for detecting stable materials and predicting their crystal structure are being debated all the time, on a range of quantitative structure-property correlations, as well as a few more ideas for using machine learning to replace first-principal methods. It looked at how active learning and surrogate-based optimization may help with the rational design process and other relevant applications. The interpretability of machine learning models and the physical understanding acquired from them are two key concerns. As a result, the many aspects of interpretability and their significance in materials research have been examined. Finally, solutions and research suggestions for a variety of computational materials science issues have been provided.

The third section of the book is based on Business Intelligence and Analytics Applications. Chapter 10, "HR ANALYTICS: Galvanizing the Organizations with the Prowess of Technology", With the emerging dependency on data for making optimized decisions, data analytics is emerging as a significant tool to increase the efficacy of the organizations. The HR department has now forayed into HR analytics to make rational decisions with less human intervention. In this tech-driven process, the data related to the workforce is assessed, evaluated, and analyzed. HR analytics is the emerging discipline adopted by organizations to give preference to analytics at all stages ranging from a candidate getting on board, to performance monitoring and stretching to scrutinizing the social media opinions of the employees. It provides information from all the touchpoints vital to employee profiling and thus leverages the competitive advantage of the organization. In the pursuit of exploring the concept and applicability of HR analytics, the authors have adopted a mixed methodology for this chapter.

In Chapter 11, "Marketing Analytics—Concept, Applications, Opportunities, and Challenges Ahead", regardless of the industry in which they operate, businesses employ a variety of marketing analytics. Technology-driven techniques and solutions are critical for optimizing marketers' connections with clients in a large and unpredictable market environment. By focusing on interaction and feedback, marketing and customer analytics have now become industry keywords as they help marketers to gain a more in-depth understanding of their consumers' experiences and make appropriate changes as a result. Analytics is used in a wide variety of fields, including consumer electronics, intelligent objects, and a rising number of other applications. This chapter provides a comprehensive overview of the rapidly growing new extent of "marketing analytics" and explores the significance of marketing analytics in practice. As a result of conducting a comprehensive analysis of studies and drawing inferences about marketing analytics, this chapter makes a significant contribution to the marketing literature. This chapter provides practitioners with an overview of recent research. Academics can use these findings to argue for and explain the value and benefits of marketing analytics in the curriculum.

Chapter 12, "Effect of Social Media Usage on Anxiety During a Pandemic: An Analytical Study on the Young Adults", social media is an integral part of everyone's daily life. Being a source of the latest information, no one is afforded to stay aloof. During the pandemic along with the information, online social networking is also on the agenda. On one hand, the usage of social media can attract social benefits which lead to increased life satisfaction, and on another, the usage of social media can also bring negative results through social overload. Against this, both the aspects of social media usage along with exploring their impact on anxiety have been investigated. The study finds that aggregate anxiety is caused by social media as a source of information and as a source of entertainment. Although the use of social media as a source of professional connection is observed as an insignificant predictor of causing anxiety.

Chapter 13, "An Exploratory Study of Understanding Consumer Buying Behavior Towards Green Cosmetics Products in the Indian Market", People's health awareness has extended from the food industry to the personal beauty industry. Consumers have increased their interest in natural ingredients, sustainable packaging, and other green elements in cosmetics. The authors have evaluated some demographic and psychographic factors that affect consumers who buy green cosmetics in the Indian region. They also identified the relationship between green consumers and social status. This chapter explores how green marketing plays a vital role in selecting buyers for green cosmetic products. The result indicated that population factors, work status, and household income significantly affect consumers who buy green beauty products. Furthermore, studies also show that green marketing does not affect consumers who buy green cosmetics.

Chapter 14, "Analytical study of Factors affecting Adoption of Blockchain by Fintech Companies", the adoption of Blockchain by fintech companies has redefined itself in many ways over the years. The investment in blockchain technology is directly linked to the various sectors that see lucrative business opportunities because of deploying distributed ledger technology. Fintech Companies have emerged as the major service providers who refine the experience of their clients by leveraging upon the knowledge pool and resources available to them. This chapter highlights that Blockchain Technology has a long way to go in the financial domain as dedicated research is on to discover the various possibilities of its application. The holistic exploratory research summarizes the various factors that determine the adoption of blockchain technology in Fintech companies. The various considerations before the adoption decision are made.

Chapter 15, "A study of the Performances of Small Cap, Large Cap and Banking & Financial Sector funds of Nippon India AMC, ICICI Prudential AMC and Tata AMC", discusses that avenues for investments in financial assets have changed drastically in the Indian capital market and mutual fund is a powerful investment choice with the potential to provide investors with long- term riches. The Indian mutual fund Industry has emerged as one of the most promising investment opportunities. Investment in financial assets has always been a matter of great importance in an investor's life. To substantiate the diversified financial goals of investors, a variety of mutual fund schemes have surfaced. It is important for both the investors and the fund managers to undergo rigorous and constant evaluation, regarding the risk & return, of various schemes under purview. It enables the fund managers to identify the strengths and weaknesses of these schemes, which helps them to take improved decisions in future. This study highlights the performance of mutual funds offered by various companies and Investors' faith in these funds, into which they put their savings from earned income.

This book will greatly benefit organizations to innovate and formulate new or improve existing strategies. As organizations are facing challenges in managing and analysing the data this book will greatly benefit them in effectively managing the voluminous data for effective decision making and problem-solving for day-to-day problems occurring in the business environment. It will benefit the decision-makers in broadening their understanding of different areas and the application of analytics in them. This book provides various applications of business analytics, so the students, researchers, academicians, practitioners, etc., get the benefit of understanding the role of business analytics in several fields.

Contents

List of Contributors

1. **Rofin TM**
National Institute of Industrial Engineering (NITIE), Mumbai, Maharashtra, 400087.

2. **Biswajit Mahanty**
Department of Industrial and Systems Engineering, Indian Institute of Technology Kharagpur, Kharagpur – 721 302, West Bengal, India.

3. **Arvind Shroff**
Indian Institute of Management, Indore.

4. **Vikas Swarnakar**
Kahlifa University, Abu Dhabi.

5. **Anthony Bagherian**
Quality Management Unit, Vectrus System Corporation, Stuttgart, Germany.

6. **Pratibha Garg**
Amity School of Business, Amity University, Uttar Pradesh, Noida.

7. **Neha Gupta**
Amity School of Business, Amity University, Uttar Pradesh, Noida.

8. **Mohini Agarwal**
Independent Researcher, Glendale, Arizona 85308 United States.

9. **Piyusha Nayyar**
Amity School of Business, Amity University, Uttar Pradesh, Noida.

10. **Salaj**
School of Basic and Applied Sciences, K R Mangalam University, Sohna Road, Gurugram-122103.

11. **Subhrata Das**
Department of operational Research, University of Delhi, Delhi-07.

12. Iosif Z Aronov

Doctor of Technical Sciences, MGIMO (Moscow State Institute of International Relations), University of the Ministry of Foreign Affairs of the Russian Federation.

13. Olga V Maksimova

Candidate of Technical Sciences, MISiS (Moscow Institute of Steel and Alloys), Federal State Budgetary Institution Yu. A. Izrael Institute of Global Climate and Ecology.

14. Mohammad Haider Syed

Department of Computer Science, College of Computing and Informatics, Saudi Electronic University, KSA.

15. Sidhu

Department of Mathematics, Shri Venkateshwara, University, Uttar Pradesh, India.

16. Kamal Upreti

Department of Information Technology, Indraprastha Engineering College, Ghaziabad, Uttar Pradesh, India.

17. Sukanta Kumar Baral

Department of Commerce, Faculty of Commerce & Management, Indira Gandhi National Tribal University (A Central University), Amarkantak, Madhya Pradesh, India.

18. Ramesh Chandra Rath

Guru Gobind Singh Educational Institutions Technical Campus, Approved by AICTE Govt. of India New Delhi and Affiliated to Jharkhand University of Technology (JUT) Ranchi, India.

19. Ruchi Jain

Amity School of Business, Amity University Uttar Pradesh, India.

20. Ruchi Khandelwal

Amity School of Business, Amity University Uttar Pradesh, India.

21. Manita Matharu

Amity School of Business, Amity University, Uttar Pradesh, Noida.

22. Amit Dangi

Faculty of Commerce and Management, SGT University, Gurugram.

23. Vijay Singh

Department of Commerce, Indira Gandhi University, Meerpur, Rewari, India.

24. Md Sohail
Jaipuria Institute of Management, Jaipur.

25. Richa Srivastava
Jaipuria Institute of Management, Lucknow.

26. Srikant Gupta
Jaipuria Institute of Management, Jaipur.

27. Pooja Mathur
Amity School of Business, Amity University, Uttar Pradesh, Noida.

28. Sony Thakural
Amity School of Business, Amity University, Uttar Pradesh, Noida.

29. Khushboo Bhasin
Amity School of Business, Amity University, Uttar Pradesh, Noida.

30. Saloni Pahuja
Amity School of Business, Amity University, Uttar Pradesh, Noida.

Section I
Operations and Supply Chain Analytics

CHAPTER 1

Wholesale Price Strategy of a Manufacturer under Collusion of Downstream Channel Members

A Game-Theoretic Approach

Rofin TM[1],* and Biswajit Mahanty[2]

Contents

1.1 Introduction

Retailers engage in collusive behavior to jointly obtain higher profits (Lin et al. 2021) leading to lower profits for the manufacturers as well as higher prices for customers (Sahuguet and Walckiers 2016). The higher prices and subsequent lower sales volume resulting from the price sensitivity

---

[1] National Institute of Industrial Engineering (NITIE), Mumbai, Maharashtra, 400087.
[2] Department of Industrial and Systems Engineering, Indian Institute of Technology Kharagpur, Kharagpur – 721 302, West Bengal, India.
* Corresponding author: rofintm@gmail.com

of the consumers are sometimes leading to a lower profit margin for the manufacturer when the downstream retailers collude. The increased margin resulting from the higher prices can compensate for the reduction in sales volume for the retailers leading to an overall higher profit when they collude. This is the rationale for retailers to collude and such behavior is reported among petrol retailers in Melbourne (ACC 2014), laundry detergent retailers (ACC 2016), and retail pharmacies (Chilet 2016). Nevertheless, the margin obtained by the manufacturer is not changing and the reduction in sales volume will hurt the profit of the manufacturer. In such a scenario, the rational manufacturer will minimize the impact of downstream collusion by adopting optimal wholesale pricing decisions. Should the manufacturer supply the products at an equal wholesale price to the downstream retailers or should he differentiate the wholesale price among the retailers?

The wholesale price strategy (WPS) of the manufacturer is governed by legislation. For instance, manufacturers complying with the Robinson-Patman Act (Yonezawa et al. 2020) or the Indian competition act, 2002 (Mazhuvanchery 2010) cannot charge different wholesale prices for downstream retailers since these acts do not permit an organization to directly or indirectly impose a discriminatory condition in the sale of goods or services. Nevertheless, wholesale price discrimination can be justified if there exists a cost disparity among the retailers owing to different methods of sale and distribution. Thus, a manufacturer, supplying the product to heterogeneous downstream retailers, can either adopt an Equal wholesale price strategy (Equal-WPS) or a Discriminatory wholesale price strategy (Dis-WPS) with the wholesale price being Equal or discriminatory respectively. Dis-WPS can be justified by the operational cost difference among the retailers. Thus, we investigate the wholesale price strategy (WPS) decision of the manufacturer considering a Dual-Channel Supply Chain (DCSC) constituted by a manufacturer and downstream brick-and-mortar store (BM-Store) and E-store when they collude.

When the manufacturer is selling the product to heterogeneous downstream channel members, the manufacturer needs to decide on impartial or discriminatory treatment. In this study, the heterogeneous channel members are a BM-Store and E-store. BM-Store and E-store are different in terms of their distribution and selling costs and this disparity can justify wholesale price discrimination from a legal point of view (Rofin and Mahanty 2021). However, the manufacturer is free to decide on whether to discriminate between the BM-Store and E-store in terms of wholesale price or not. If the wholesale price discrimination yields higher profit for the manufacturer under downstream collusion, he will adopt a Dis-WPS, or else Equal-WPS, despite the operational cost differences between the downstream chain members. With this background, the following research objectives have been addressed in this chapter.

- To assess the effect of wholesale price discrimination on the performance and pricing decisions of the chain members (BM-Store, E-store, and manufacturer) when BM-Store and E-store collude.
- To ascertain the performance variation of the channel members based on product-web fit.

To address the above-mentioned research objectives, game models have been designed. The remaining content of the chapter is structured as follows. The literature review is reported in Section 2. Section 3 deals with the model. Section 4 presents the derivation of profit-maximizing decisions and optimal profit of the channel members using equilibrium analysis. In Section 5, the results are reported based on a numerical example. Discussions and Managerial Implications are given in Section 6. Section 7 deals with Conclusions and Directions for Future Research.

1.2 Related Literature

The research papers related to this study are primarily the works on the WPS of the manufacturer or upstream supply chain (SC) member supplying products to downstream competing retailers. In the past literature, several issues have been addressed within this SC structure. For instance, cooperative advertisement (Aust and Buscher 2014, Karray and Amin 2015), revenue sharing contract (Yao et al. 2008, Pan et al. 2010), channel power (Wang et al. 2011, Jafari et al. 2016), return policies (Savaskan and Van Wassenhove 2006, Parthasarathi et al. 2011, Chen and Grewal 2013). There are very few studies (Ali et al. 2018, De Giovanni 2018) that have considered Dis-WPS.

It can be observed that most of these works have employed game-theoretic models with Parthasarathi et al. (2011) employing a combination of a newsvendor model along with game theory whereas Xu et al. (2013) employed simulation in addition to game theory. De Giovanni (2018) applied a dynamic game model and Ma and Xie (2016a) carried out a complexity analysis. Recently, the DCSC literature has shifted its focus to contemporary issues like greening strategies, showrooming, and fulfillment strategies. For instance, Chai et al. (2021) applied the Hotelling game theory model for a DCSC and established store brand as a mechanism to combat showrooming. Qiu et al. (2021) studied the competition between a retailer and an E-tailer under the DCSC structure for two fulfillment policies of the E-tailer, i.e., (i) drop shipping and (ii) order batching. They considered demand uncertainty in their model and found that the preference of the channel members for the fulfillment policy varies according to the magnitude of market share and profit-sharing ratio. Shao (2021) modeled a DCSC in which the manufacturer sells through competing retailers and examined the impact of the omnichannel move of the retailers on supply chain performance. They found that omnichannel retailing is not desirable for the manufacturer. Yang et al. (2021) investigated the effect of online consumer reviews in a DCSC where the online channel is a direct channel. They reported conditions under which the retail price and online price will be varied based on online consumer reviews. Zhang et al. (2021) examined the dynamic pricing and greening strategies of channel members in a DCSC under a two-period model for both centralized and decentralized settings. They found that a cost-sharing contract is effective in improving SC performance. Though numerous studies are present addressing various aspects of SC management under a DCSC configuration, there is scant literature addressing the WPS of the manufacturer when the heterogeneous downstream BM-Stores are engaged in collusion. Most of the studies have considered Equal-WPS despite the option of selling products at different wholesale prices to downstream channel members. Further, the possibility of collusion between the channel members remains unexplored in DCSC literature. We bridge this gap by investigating the impact of the WPS of a manufacturer on the performance of channel members by modeling a DCSC where the downstream channel members are engaged in collusion.

1.3 Research Model

This section explains the research model. A two-echelon DCSC has been considered with BM-Store and E-store engaged in collusion as shown in Fig. 1.

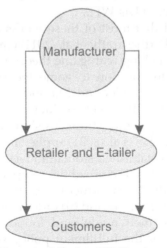

Figure 1: Collusion (CL) Structure.

Demand for the BM-Store and E-store is as follows

$$D_r = a(1-\theta) - \lambda p_r + \gamma p_e \tag{1}$$

$$D_e = a\theta - \lambda p_e + \gamma p_r \tag{2}$$

Where, a is the base demand, and θ is the product-web fit (PWF). PWF is a measure of the suitability of a product to be sold through an online platform that varies across different product categories (Hong and Pavlou 2014). Some products are more suitable to be sold through online platforms such as books, e-books, software, etc., since a pre-purchase evaluation is not very critical in such products. Certain products need to be evaluated for their features and characteristics through a direct experience such as shoes, apparel, jewelry, LED TVs, etc. For such products the value of PWF is low. λ and γ are own-price elasticity and cross-price elasticity respectively where $\lambda > \gamma$. (Ding et al. 2016). p_r and p_e are the prices of the BM-Store and E-store respectively. w represents the wholesale price under Equal-WPS. Under Dis-WPS, w_r and w_e indicate the wholesale price for the BM-Store and E-store respectively. Cr and Ce denote the marginal cost of distribution of the BM-Store and E-store respectively. In this model, a Stackelberg game (Dan et al. 2012) is modeled on account of the channel leadership of the manufacturer.

1.4 Equilibrium Analysis of Collusion (CL) Game

This section reports the equilibrium analysis of the collusion game for (i) Equal-WPS and (ii) Dis-WPS.

Case 1: Equal-WPS: In this game, first, the manufacturer announces his w. After observing the w BM-Store and E-store choose p_r and p_e respectively. The backward induction procedure used to solve the Stackelberg game is as follows.
The demand functions of the BM-Store and E-store are

$$D_{r1}^{CL} = a(1-\theta) - \lambda p_{r1}^{CL} + \gamma p_{e1}^{CL} \tag{3}$$

$$D_{e1}^{CL} = a\theta - \lambda p_{e1}^{CL} + \gamma p_{r1}^{CL} \tag{4}$$

The profit functions of the BM-Store and E-store are

$$\pi_{r1}^{CL} = \left(p_{r1}^{CL} - w^{CL} - c_r\right) D_{r1}^{CL} = \left(p_{r1}^{CL} - w^{CL} - c_r\right)\left(a(1-\theta) - \lambda p_{r1}^{CL} + \gamma p_{e1}^{CL}\right) \tag{5}$$

$$\pi_{e1}^{CL} = \left(p_{e1}^{CL} - w^{CL} - c_e\right) D_{e1}^{CL} = \left(p_{e1}^{CL} - w^{CL} - c_e\right)\left(a\theta - \lambda p_{e1}^{CL} + \gamma p_{r1}^{CL}\right) \tag{6}$$

The total profit of the BM-Store and E-store is

$$\pi_T = \pi_{r1}^{CL} + \pi_{e1}^{CL} \tag{7}$$

The concavity of π_T with respect to p_{r1}^{CL} and p_{e1}^{CL} is proved as given below

$$H(\pi_{m2}^{HN}) = \begin{bmatrix} -2\lambda & 2\gamma \\ 2\gamma & -2\lambda \end{bmatrix} \tag{8}$$

In the Hessian, $H_{11} < 0$, $H_{22} < 0$ and $H_{11}H_{22} - (H_{12})^2 > 0$. Therefore, the profit-maximizing values p_{r1}^{CL} and p_{e1}^{CL} can be obtained by solving the respective FOCs π_T as shown below.

Theorem 1: *Under Equal-WPS, the profit-maximizing price for the BM-Store and E-store under collusion, are*

$$p_{r1}^{CL} = \frac{a(1-\theta) + w^{CL}(\lambda - \gamma) - \gamma c_e + \lambda c_r + 2\gamma p_e^{CL}}{2\lambda} \tag{9}$$

$$p_{e1}^{CL} = \frac{a\theta + w^{CL}(\lambda - \gamma) + \lambda c_e - \gamma c_r + 2\gamma p_r^{CL}}{2\lambda} \tag{10}$$

Substitute p_{r1}^{CL*} and p_{e1}^{CL*} into Eqn. (3) and Eqn. (4) to obtain the profit-maximizing sales volume of the BM-Store and E-store as reported below.

Corollary 1: *Under* Equal-WPS, *the profit-maximizing sales volume of the BM-Store and E-store, under collusion, are*

$$Q_{r1}^{CL} = \frac{1}{2}\left(a(1-\theta) - w^{CL}(\lambda - \gamma) + \gamma c_e - \lambda c_r\right) \tag{11}$$

$$Q_{e1}^{CL} = \frac{1}{2}\left(a\theta - w^{CL}(\lambda - \gamma) - \lambda c_e + \gamma c_r\right) \tag{12}$$

Anticipating the profit-maximizing decisions of the BM-Store and E-store, the rational manufacturer decides on his profit-maximizing wholesale price.

$$\pi_{m1}^{CL} = (w^{CL} - s)\left(Q_{r1}^{CL*} + Q_{e1}^{CL*}\right) \tag{13}$$

Since π_{m1}^{CL} is concave w^{CL} as in other cases, the profit-maximizing wholesale price can be derived from the FOC of π_{m1}^{CL} with respect to w^{CL}.

Theorem 2: *Under Equal-WPS, the profit-maximizing wholesale price of the manufacturer, is*

$$w^{CL} = \frac{a + (\lambda - \gamma)(2s - c_e - c_r)}{4(\lambda - \gamma)} \tag{14}$$

Substitute the profit-maximizing decisions into Eqn. (5), Eqn. (6) and Eqn. (13), to derive the optimal profit of the BM-Store, E-store, and manufacturer as reported below.

Corollary 2: *Under Equal-WPS, the optimal profit of the BM-Store, E-store, and manufacturer are*

$$\pi_{r1}^{CL*} = \frac{\left(\left(\lambda^2 - \gamma^2\right)(2s + 3c_r - c_e) - a\left(\gamma(4\theta - 1) + \lambda(3 - 4\theta)\right)\right)}{64\left(\lambda^2 - \gamma^2\right)} \tag{15}$$

$$\pi_{e1}^{CL*} = \frac{\left(\left(\lambda^2 - \gamma^2\right)(c_r - 2s - 3c_e) - a\left(\gamma(4\theta - 3) + \lambda(1 - 4\theta)\right)\right)}{64\left(\lambda^2 - \gamma^2\right)} \tag{16}$$

$$\pi_{m1}^{CL*} = \frac{\left(a - 2s(\lambda - \gamma) - (\lambda - \gamma)(c_e + c_r)\right)^2}{16(\lambda - \gamma)} \tag{17}$$

Case 2: Dis-WPS: When the manufacturer follows Dis-WPS, the sequence of the game is the same as in Case 1. Therefore, backward induction is applied to find the profit-maximizing decisions of channel members.

The demand functions of the BM-Store and E-store are

$$D_{r2}^{CL} = a(1 - \theta) - \lambda p_{r2}^{CL} + \gamma p_{e2}^{CL} \tag{18}$$

$$D_{e2}^{CL} = a\theta - \lambda p_{e2}^{CL} + \gamma p_{r2}^{CL} \tag{19}$$

The profit functions of the BM-Store and E-store are

$$\pi_{r2}^{CL} = \left(p_{r2}^{CL} - w_{r2}^{CL} - c_r\right)D_{r2}^{CL} = \left(p_{r2}^{CL} - w_{r2}^{CL} - c_r\right)\left(a(1 - \theta) - \lambda p_{r2}^{CL} + \gamma p_{e2}^{CL}\right) \tag{20}$$

$$\pi_{e2}^{CL} = \left(p_{e2}^{CL} - w_{e2}^{CL} - c_e\right)D_{e2}^{CL} = \left(p_{e2}^{CL} - w_{e2}^{CL} - c_e\right)\left(a\theta - \lambda p_{e2}^{CL} + \gamma p_{r2}^{CL}\right) \tag{21}$$

The total profit function of the BM-Store and E-store is

$$\pi_T = \pi_{r2}^{CL} + \pi_{e2}^{CL} \tag{22}$$

Since π_T is concave in p_{r2}^{CL} and p_{e2}^{CL}, the profit-maximizing values p_{r2}^{CL} and p_{e2}^{CL} can be obtained by solving the respective FOCs π_T and the results are reported below.

Theorem 3: *The profit-maximizing price for the BM-Store and E-store under Dis-WPS, are*

$$p_{r2}^{CL*} = \frac{a(1-\theta) - \gamma c_e + \lambda c_r + 2\gamma p_e^{CL} - \gamma w_e^{CL} + \lambda w_r^{CL}}{2\lambda} \tag{23}$$

$$p_{e2}^{CL*} = \frac{a\theta + \lambda c_e - \gamma c_r + 2\gamma p_r^{CL} + \lambda w_e^{CL} - \gamma w_r^{CL}}{2\lambda} \tag{24}$$

Substitute p_{r1}^{CL*} and p_{e1}^{CL*} into Eqn. (18) and Eqn. (19) to obtain the optimal sales volume of BM-Store and E-store as reported below.

Corollary 3: *The profit-maximizing sales volume of the BM-Store and E-store, under Dis-WPS, are*

$$Q_{r2}^{CL*} = \frac{1}{2}\left(a(1-\theta) + \gamma\left(c_e + w_{e2}^{CL}\right) - \lambda\left(c_r + w_{r2}^{CL}\right)\right) \tag{25}$$

$$Q_{e2}^{CL*} = \frac{1}{2}\left(a\theta - \lambda\left(c_e + w_{e2}^{CL}\right) + \gamma\left(c_r + w_{r2}^{CL}\right)\right) \tag{26}$$

Now, the manufacturer decides on his profit-maximizing wholesale price.

$$\pi_{m2}^{CL} = (w_{r2}^{CL} - s)Q_{r2}^{CL*} + (w_{e2}^{CL} - s)Q_{e2}^{CL*} \tag{27}$$

Since π_{m2}^{CL} is concave in w_{r2}^{CL} and w_{r2}^{CL}, the profit-maximizing values w_{e2}^{CL} and w_{e2}^{CL} can be obtained from the FOC of π_{m2}^{CL}

Theorem 4: *The profit-maximizing wholesale price for the BM-Store and E-store under Dis-WPS are*

$$w_{r2}^{CL*} = \frac{a\gamma\theta + a\lambda(1-\theta) + (s-c_r)\left(\lambda^2 - \gamma^2\right)}{2\left(\lambda^2 - \gamma^2\right)} \tag{28}$$

$$w_{e2}^{CL*} = \frac{a\gamma(1-\theta) + a\theta\lambda + (s-c_e)\left(\lambda^2 - \gamma^2\right)}{2\left(\lambda^2 - \gamma^2\right)} \tag{29}$$

Substitute the profit-maximizing decisions into Eqn. (20), Eqn. (21) and Eqn. (27), to obtain the optimal profit of the channel members as reported below.

Corollary 4: *The optimal profit of the BM-Store, E-store, and manufacturer under Dis-WPS are*

$$\pi_{r2}^{CL*} = \frac{\left(a(1-\theta) - s(\lambda - \gamma) + \gamma c_e - \lambda c_r\right)\left(a(\gamma\theta + \lambda(1-\theta)) - (\lambda^2 - \gamma^2)(s + c_r)\right)}{16\left(\lambda^2 - \gamma^2\right)} \tag{30}$$

$$\pi_{e2}^{CL*} = \frac{\left((s+c_e)(\lambda^2 - \gamma^2) - a(\gamma(1-\theta) + \theta\lambda)\right)\left(s(\lambda - \gamma) - a\theta + \lambda c_e - \gamma c_r\right)}{16(\lambda^2 - \gamma^2)} \quad (31)$$

$$\pi_{m2}^{CL*} = \frac{(2s^2(\gamma - \lambda)^2(\gamma + \lambda) - a^2(2(1-\theta)\theta(\lambda - \gamma) - \lambda) - 2as(\lambda^2 - \gamma^2) - K)}{8(\lambda^2 - \gamma^2)} \quad (32)$$

$$K = (\lambda^2 - \gamma^2)\left(c_e(2s(\gamma - \lambda) + 2a\theta - \lambda c_e) + 2(a(1-\theta) + s(\gamma - \lambda) + \gamma c_e)c_r - \lambda c_r^2\right)$$

The complex equations of profit-maximizing decisions and resulting profit make it difficult to compare them to understand the performance of channel members. Therefore, we resort to a numerical example.

1.5 Results and Discussion Through Numerical Experiment

Different numerical experiments were conducted, and a common pattern of results was obtained across the different numerical sets. Therefore, a representative instance is reported with the following values.

$a = 300$; $\lambda = 1.5$; $\gamma = 1.2$; $c_r = 7$; $c_e = 10$; $s = 50$

We carry out a sensitivity analysis of the profit-maximizing decisions and corresponding profit of the channel members with respect to PWF (θ). The rationale for selecting θ is its product-wise difference (Kacen et al. 2013). Further, θ is rapidly changing especially in developing countries with rising internet access and usage. The customers are increasingly relying on e-shopping thereby causing a positive shift in product-web fit. Therefore, we carry out a sensitivity analysis regarding product-web fit (θ).

First, the sensitivity of wholesale price is examined and the results of wholesale price comparison for Equal-WPS and Dis-WPS are presented in Table 1.

Table 1 reports that (i) Under Equal-WPS, w is independent of the PWF for the product. (ii) Under Dis-WPS, the w decreases with an increase in PWF, and the w_e increases with an increase in PWF. With the increase in PWF, the base demand for BM-Store decreases, and that of the E-store increases. This is the rationale behind the manufacturer charging higher w_e and w_r lower with the increase in PWF. Now we examine the price sensitivity of the BM-Store and E-store.

Table 2 reports that (i) The p_r decreases with an increase in PWF irrespective of the WPS. (iii) p_r is higher under Dis-WPS compared to that under Equal-WPS in the case of products with Low-PWF and it is lower under Dis-WPS compared to that under Equal-WPS in the case of products

Table 1: Wholesale Price vs. PWF.

PWF (θ)	w_r CL Equal-WPS	w_r CL Dis-WPS	w_e CL Equal -WPS	w_e CL Dis-WPS
0.05	275.00	300.00	275.00	250.00
0.15	275.00	294.44	275.00	255.56
0.25	275.00	288.88	275.00	261.11
0.35	275.00	283.33	275.00	266.67
0.45	275.00	277.77	275.00	272.22
0.55	275.00	272.22	275.00	277.78
0.65	275.00	266.66	275.00	283.33
0.75	275.00	261.11	275.00	288.89
0.85	275.00	255.55	275.00	294.44
0.95	275.00	250.00	275.00	300.00

Table 2: Price vs. PWF.

PWF (θ)	p_r CL Equal-WPS	p_r CL Dis-WPS	p_e CL Equal-WPS	p_e CL Dis-WPS
0.05	412.50	425.00	362.50	350.00
0.15	406.94	416.67	368.06	358.33
0.25	401.39	408.33	373.61	366.67
0.35	395.83	400.00	379.17	375.00
0.45	390.28	391.67	384.72	383.33
0.55	384.72	383.33	390.28	391.67
0.65	379.17	375.00	395.83	400.00
0.75	373.61	366.67	401.39	408.33
0.85	368.06	358.33	406.94	416.67
0.95	362.50	350.00	412.50	425.00

with High-PWF. (i) p_e increases with an increase in PWF irrespective of the WPS. (iii) p_e is lower under Dis-WPS compared to that under Equal-WPS in the case of products with Low-PWF and it is higher under Dis-WPS compared to that under Equal-WPS for products with High-PWF. The variation in the base demand with respect to variation in PWF explains the dynamics of price variation. Whenever there is an increase in the base demand, it will increase in price and vice versa. Nevertheless, our focus is on how the dynamics vary with respect to the WPS, and price variation is contingent on the WPS. Now we examine the sensitivity in the sales volume and the results are reported in Table 3.

Table 3: Sales Volume vs. PWF.

PWF (θ)	Q_r CL Equal -WPS	Q_r CL Dis-WPS	Q_e CL Equal-WPS	Q_e CL Dis-WPS
0.05	101.25	67.50	0.00	0.00
0.15	86.25	60.00	0.00	7.50
0.25	71.25	52.50	0.00	15.00
0.35	56.25	45.00	11.25	22.50
0.45	41.25	37.50	26.25	30.00
0.55	26.25	30.00	41.25	37.50
0.65	11.25	22.50	56.25	45.00
0.75	0.00	15.00	71.25	52.50
0.85	0.00	7.50	86.25	60.00
0.95	0.00	0.00	101.25	67.50

Table 3 reports that: (i) The Q_r decreases with an increase in PWF of the product irrespective of the WPS, and (ii) The Q_r is higher under Equal-WPS compared to that of Dis-WPS for products with Low-PWF whereas the Q_r is lesser under Dis-WPS compared to that of Equal-WPS for products with High-PWF. (i) Q_e increases with an increase in PWF of the product irrespective of the WPS, and (ii) The Q_e is higher under Dis-WPS compared to that of Equal-WPS for products with Low-PWF whereas Q_e is lesser under Dis-WPS compared to that of Equal-WPS for products with High-PWF. Next, we examine the profit of the SC members.

1.5.1 Profit of the BM-Store

Figure 2 shows the profit of the BM-Store under Equal-WPS and Dis-WPS.

From Fig. 2, it can be stated that (i) Irrespective of the WPS, the π_r^{CL} decreases with an increase in PWF. (ii), the BM-Store enjoys higher profit under Equal-WPS compared to the profit under Dis-WPS for products with Low-PWF. For High-PWF products, Dis-WPS yields a higher profit compared to Equal-WPS for the BM-Store. (iii) There is a threshold value of PWF above which the profit of the BM-Store is greater under Dis-WPS. In the case of our numerical example, the threshold value is found to be 0.52. There can be variations in the threshold value based on the numerical example. Nevertheless, the overall behavior of the profit variation with respect to PWF remains the same. Next, we report the variation of the profit of the E-store with respect to PWF in the case of Equal-WPS and Dis-WPS.

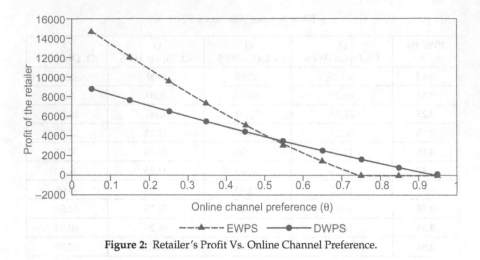

Figure 2: Retailer's Profit Vs. Online Channel Preference.

1.5.2 *Profit of the E-store*

Figure 3 shows the profit of the E-store under Equal-WPS and Dis-WPS.

From Fig. 3, it can be noted that: (i) Irrespective of WPS, π_e^{CL} increases with an increase in PWF for the product. (ii) E-store enjoys higher profit under Equal-WPS compared to the profit under Dis-WPS for products with High-PWF. (iii) For Low-PWF products, Dis-WPS yields higher profits compared to Equal-WPS for the BM-Store. (iv) There is a threshold value of PWF above which π_e^{CL} is lower under Dis-WPS. As in the case of the threshold value for the BM Store, the threshold value for the E-store

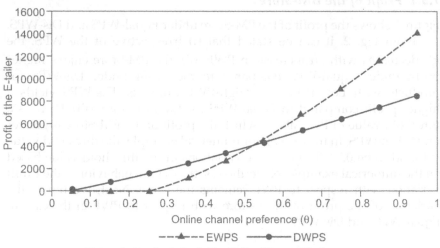

Figure 3: E-tailer's Profit Vs. Online Channel Preference.

is found to be 0.52. Thus, E-store's case is a complementary case to that of the BM-Store's case.

1.5.3 Profit of the Manufacturer

The profit of the manufacturer is plotted in Fig. 4.

From Fig. 4, the following observations are made (i) For products with Low-PWF and High-PWF, the manufacturer should adopt Equal-WPS to obtain greater profit when the BM-Store and E-store collude. (ii) For products with Moderate-PWF the difference in profit between Equal-WPS and Dis-WPS seems to be minimal from Fig. 4. However, when aggregated over the values corresponding to the moderate-PWF, it was found that Dis-WPS yields higher profit for the manufacturer. Therefore, the manufacturer prefers Dis-WPS for products with Moderate-PWF when the downstream BM-Store and E-store collude.

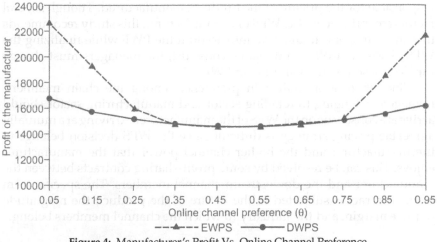
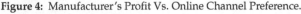

Figure 4: Manufacturer's Profit Vs. Online Channel Preference.

1.6 Discussions and Managerial Implications

This study has several implications for practicing managers representing both the manufacturing sector and retailing sector. The recommendations for managers can be obtained from the numerical example employed in the study. From the numerical example, the preference of chain members for different product categories has been identified for Equal-WPS and Dis-WPS based on the average value of profit as summarized in Table 4.

From Table 4, it can be observed that there is a clear conflict in preference between the BM-Store and E-store concerning the WPS. Nevertheless, the conflict in preference between the BM-Store and manufacturer and

Table 4: Wholesale Price Strategy Preference of the Chain Members.

	BM-Store	E-store	Manufacturer
Low-PWF Products	Equal-WPS	Dis-WPS	Equal -WPS
Moderate-PWF Products	Equal -WPS	Dis-WPS	Dis-WPS
High-PWF Products	Dis-WPS	Equal -WPS	Equal -WPS

that between E-store and manufacturer are partial. BM-Store's preference and manufacturer's preference match in the case of Low-PWF products. Similarly, the E-store's preference and manufacturer's preference match in the case of both Moderate-PWF products and High-PWF products. In this study, we have assumed that the manufacturer holds higher channel power relative to the other chain members. Therefore, the Low-PWF and High-PWF products manufacturer will take Equal-WPS and the High-PWF manufacturer will take Dis-WPS. This finding is helpful for the managers responsible for the pricing of the products manufactured. Though several factors are influencing the WPS of a manufacturer, this study recommends that the managers should take into account the PWF while finalizing the WPS. Since the PWF is a dynamic construct, the managers must change the WPS based on the variation in PWF.

Since there is a conflict in preference among the chain members, managers belonging to retailing sector and manufacturing sector should arrive at a consensus on the WPS of the manufacturer. Arriving at a mutually agreeable pricing strategy is difficult since the WPS decision belongs to the manufacturer and the higher channel power that the manufacturer enjoys. This can be resolved by some profit-sharing contracts between the manufacturer and the downstream channel members. The specific form of the contract is subjected to the nature of the product, the magnitude of price margin, and the industry to which the channel members belong.

1.7 Conclusions and Directions for Future Research

In this chapter, the impact of the wholesale pricing strategy of the manufacturer is investigated when the downstream chain is comprised of the colluding BM-Store and an E-store. Game-theoretic models have been developed corresponding to two wholesale pricing strategies of the manufacturer, i.e., (i) Equal Wholesale Pricing Strategy, and (ii) Discriminatory Wholesale Pricing Strategy. Equilibrium analysis was carried out to derive the optimal wholesale price, optimal price, optimal sales volume, and optimal profit of the supply chain members. The game-theoretic models are compared through a numerical example capturing the difference between product categories based on their product-web fit.

This study can be extended in several directions. The conflict in preference observed among chain members warrants research on the alternate channel power structures such as BM-Store leader or E-store leader supply chains. Further, the study can be extended by considering uncertain demand which can make the study closer to reality. During the COVID-19 pandemic, there were several demand disruptions. Some of them were positive demand disruptions and some were negative demand disruptions. It will be worthwhile to model these demand disruptions and their impact on the performance of the channel members under a collusive setting and a non-collusive setting.

References

Ali, S. M., Md. H. Rahman, T. J. Tumpa, A. A. M.l Rifat, and S. K. Paul. 2018. Examining price and service competition among BM-Stores in a supply chain under potential demand disruption. *Journal of Retailing and Consumer Services* 40: 40–47.

Aust, G., and U. Buscher. 2014. Vertical cooperative advertising in a BM-Store duopoly. *Computers & Industrial Engineering* 72: 247–254.

Australian competition & consumer commission (ACCC). 2014 ACCC takes action against Informed Sources and petrol BM-Stores for price information sharing. https://www.accc.gov.au/media-release/accc-takes-action-against-informed-sources-and-petrol-BM-Stores-for-price-information-sharing

Australian competition & consumer commission (ACCC). 2016. Woolworths was ordered to pay $9 million in penalties in laundry detergent cartel proceedings. https://www.accc.gov.au/media-release/woolworths-ordered-to-pay-9-million-in-penalties-in-laundry-detergent-cartel-proceedings.

Chai, L., D. D. Wu, A. Dolgui, and Y. Duan. 2021. Pricing strategy for BM-Store in a dual-channel supply chain based on hotelling model. *International Journal of Production Research* 59(18): 5578–5591.

Chen, J., and R. Grewal. 2013."Competing in a supply chain via full-refund and no-refund customer returns policies." *International Journal of Production Economics* 146(1): 246–258.

Chilet, Jorge Alé. 2018. Gradually rebuilding a relationship: The emergence of collusion in retail pharmacies in Chile.

Dan, B., G. Xu, and C. Liu. 2012. Pricing policies in a dual-channel supply chain with retail services. *International Journal of Production Economics* 139(1): 312–320.

De Giovanni, P. 2018. A joint maximization incentive in closed-loop supply chains with competing BM-Stores: The case of spent-battery recycling. *European Journal of Operational Research* 268(1): 128–147.

Ding, Q., C. Dong, and Z. Pan. 2016. A hierarchical pricing decision process on a dual-channel problem with one manufacturer and one BM-Store. *International Journal of Production Economics* 175: 197–212.

Hong, Y., and P. A. Pavlou. 2014. Product fit uncertainty in online markets: Nature, effects, and antecedents. *Information Systems Research* 25(2): 328–344.

Jafari, H., S. R. Hejazi, and M. Rasti-Barzoki. 2016.Pricing decisions in dual-channel supply chain including monopolistic manufacturer and duopolistic BM-Stores: a game-theoretic approach. *Journal of Industry, Competition and Trade* 16(3): 323–343.

Kacen, J. J., J. D. Hess, and W-Y. K. Chiang. 2013. Bricks or clicks? Consumer attitudes toward traditional stores and online stores. *Global Economics and Management Review* 18(1): 12–21.

Karray, S., and S. H. Amin. 2015. Cooperative advertising in a supply chain with retail competition. *International Journal of Production Research* 53(1): 88–105.

Lin, J., X. Ma, S. Talluri, and C-H. Yang. 2021. Retail channel management decisions under collusion. *European Journal of Operational Research*.

Ma, J., and L. Xie. 2016a. Study on the complexity pricing game and coordination of the duopoly air conditioner market with disturbance demand. *Communications in Nonlinear Science and Numerical Simulation* 32: 99–113.

Mazhuvanchery, S.V. 2010. The Indian competition act: a historical and developmental perspective. *Law and Development Review* 3(2): 241–270.

Pan, K., K. K. Lai, S. C. H. Leung, and D. Xiao. 2010. Revenue-sharing versus wholesale price mechanisms under different channel power structures. *European Journal of Operational Research* 203(2): 532–538.

Parthasarathi, G., S. P. Sarmah, and M. Jenamani. 2011. Supply chain coordination under retail competition using stock dependent price-setting newsvendor framework. *Operational Research* 11(3): 259–279.

Qiu, R., L. Hou, Y. Sun, M. Sun, and Y. Sun. 2021. Joint pricing, ordering and order fulfillment decisions for a dual-channel supply chain with demand uncertainties: A distribution-free approach. *Computers & Industrial Engineering* 160: 107546.

Rofin, T. M., and B. Mahanty. 2021. Impact of wholesale price discrimination on the profit of chain members under different channel power structures. *Journal of Revenue and Pricing Management* 91–107.

Sahuguet, N., and A. Walckiers. 2017. A theory of hub-and-spoke collusion. *International Journal of Industrial Organization* 53: 353–370.

Savaskan, R. C., and L. N. Van Wassenhove. 2006. Reverse channel design: the case of competing BM-Stores. *Management science* 52(1): 1–14.

Shao, X. 2021. Omnichannel retail move in a dual-channel supply chain. *European Journal of Operational Research* 294(3): 936–950.

Wang, S.-D., Y.-W. Zhou, J. Min, and Y.-G. Zhong. 2011. Coordination of cooperative advertising models in a one-manufacturer two-BM-Store supply chain system. *Computers & Industrial Engineering* 61(4): 1053–1071.

Xu, Q., Z. Liu, and B. Shen. 2013. The impact of price comparison service on pricing strategy in a dual-channel supply chain. *Mathematical Problems in Engineering*.

Yao, Z., S. C. H. Leung, and K. K. Lai. 2008. Manufacturer's revenue-sharing contract and retail competition. *European Journal of Operational Research* 186(2): 637–651.

Yang, W., J. Zhang, and H. Yan. 2021. Impacts of online consumer reviews on a dual-channel supply chain. *Omega* 101: 102266.

Yonezawa, K., M. I. Gómez, and T. J. Richards. 2020. The robinson–patman act and vertical relationships. *American Journal of Agricultural Economics*, 102(1): 329–352.

Zhang, C., Y. Liu, and G. Han. 2021. Two-stage pricing strategies of a dual-channel supply chain considering public green preference. *Computers & Industrial Engineering* 151: 106988.

CHAPTER 2

AI and ML in Supply Chain Decision Making

A Pragmatic Discussion

Arvind Shroff

Contents

2.1 Introduction

In the current scenario of the ever-increasing competitive market, efficient supply chain execution is one of the profit-making pillars of every organization. It is a function of how smoothly the uncertainty between the demand and supply is matched, leading to on-time fulfillment of customer needs. Demand Profiling is an essential strategic tool used in the predictive analytics domain of supply chain management to handle demand-supply mismatch and the uncertainty arising from it. Demand profiling is the primary driver of supply chain strategy at an SKU (Stock Keeping Unit) level (using the 4Vs of demand classifiers viz volume,

PhD Candidate, Operations Management & Quantitative Techniques Area, Indian Institute of Management, Indore. Email: f18arvinds@iimidr.ac.in variety, variation, and visibility). Demand profiling is based on AI and ML capabilities and can be used for efficient supply planning, demand planning, and forecasting at the component level and helps in inventory and logistics profiling. The supply chain's complexity depends on customer demand profile, geographic locations, and lead time.[1] With the help of Artificial Intelligence (AI) and Machine Learning (ML), we can turn data into valuable and executable insights giving us information about demand-based segmentation in supply chains. Demand profiled using AI capabilities is often considered a modular and tactical tool that addresses the problems related to complex supply chains such as longer lead times and high inventory levels.

The health of the Supply Chain (SC) depends upon the organization's ability to horizontally and vertically integrate the end-to-end processes of procuring raw materials, transforming them into finished products, and delivering them to the customers. Many leading organizations have expressed *real-time invisibility* and *information asymmetry* across various levels in the entire chain as hindrances to achieving excellence. Lately, supply chain experts have focused on applying data-intensive insights on the various functional aspects of the chain-like inventory, warehouses, logistics, and transportation. This has opened avenues for increasing profitability through informed decision-making with the help of structured data analysis. AI and ML are some of the ways that have been around for decades now but have been underutilized in terms of decision-making in the context of Supply Chain Management (SCM).

2.2 Literature Review

Čerka et al. (2015) defined AI as the ability of a system to achieve a specific goal by replicating human intelligence and making rational decisions. Min (1995) explored the potential benefits of various AI techniques in solving the relevant problems of SCM and reviewed the successful case studies of AI applications. AI-enabled machines possess the ability to self-learn through the data, impute missing data by extrapolating the available data, and gain insights through proper reasoning (Samuel 1995). In general, AI is referred to as the utilization of computers for perceiving examples, learning from certain behaviors, obtaining knowledge, and exploring different types of inferences to take care of issues in decision-making under circumstances where optimal or ideal solutions are costly or hard to deliver (Russell and Norvig 1995, Luger 2002). By using state-of-the-art

---

[1] **Source:** Accessed on June 21, 2021, from the article *Demand Profiling – A small but pragmatic approach in Data Analytics*, https://3scanalyticseu.blogspot.com/2019/04/demand-profiling-small-but-pragmatic.html

algorithms, AI possesses the ability to comprehend SC data and decipher the various areas of development in SC processes (Ni et al., 2020). Very recently, Singh et al. (2022) explored the emerging applications of AI in SCM and the associated challenges through a review study. They found that AI can optimize operations at various levels of the supply chain to improve efficiency with minimum cost. In essence, AI possesses the power to comprehend human decision-making skills and replicate this phenomenon with the help of computer systems for problem-solving.

AI and ML in the supply chain are full of potential and innovation, which remains unexplored. Min (2010) is regarded as the first review work to focus on the impact of seven AI tools on eight SCM activities through a corpus of 28 articles. The review revealed the widespread acceptance of AI as a tool for aiding human decision-making due to its ability to analyze business data intelligently by accurately learning the business phenomena. The next was an industry-specific review undertaken by Ngai et al. (2014), who collected 77 articles from 1994 and explored the applications of seven AI techniques in the textile and apparel industry. They found that gaps between AI techniques and SCM decision-making have led to the emergence of ML techniques like Artificial Neural Networks (ANN), as confirmed by Ni et al. (2019). The academicians have attempted to review the managerial aspect of AI application in SC, but the work related to the process innovation created by AI at different levels of the SC remains scarce. Riahi et al. (2021) conducted a systematic literature review focusing on the process perspective to fill in this gap. They classified 136 scholarly articles published between 1996 and 2020 through four structural dimensions—level of analytics, algorithms, industry, and supply chain processes. Our work extends this work by examining the application of various techniques of AI and ML across the sectors of practitioner-driven SC decision-making such as warehouse management, demand planning, and inventory management. Although practitioners have found the utility of AI in such aspects, the current research work still has the lacunae to identify the potential leverage of AI in SCM. Accordingly, to fill in the knowledge gap and serve as a reference for future researchers, we aim to address the following broad objectives through this study:

RO 1: To identify the various fields of AI in terms of their applications in the efficient management of supply chains.

RO 2: To understand the existing literature relating to applications of AI in SCM and explore their benefits.

RO 3: To establish the present research scope in AI and earmark the potential AI-based SCM solutions that can be explored.

RO 4: To discuss the future implications and extensions of AI when diversified onto the emerging fields of the supply chain.

2.3 AI Technologies for Supply Chain Planning

AI can be portrayed as a catch-all term for innovation that includes intelligent machines trained to act and reason like humans. Furthermore, keeping in mind that a significant number of those fields do fall under the AI umbrella, it is still vital to know the key differences between the terminologies. Here are three straightforward meanings of AI and related innovations and standard definitions on how they apply to supply chain planning.

2.3.1 Artificial Intelligence

The inbuilt capability of a machine to perform human-like functions, for instance, learning, cooperating with the surroundings, critical thinking, and practicing creative decision making to make choices, execute the range of plans and accomplish desired objectives. The applications of AI in supply chain planning include:

- Increases the value created by supply chain planners who, based on analytical tools of AI, analyze historical and real-time data and recommend production planning and dispatch planning, which helps bridge the gap between demand and supply of products in the chain.
- Improves the accuracy of predicting risk and disruptions in the supply chain, working on data from multiple sources like stock market data, news, social media, etc.

2.3.2 Machine Learning

A significant application of AI, machine learning, coined by Samuel (1995), is a method to investigate how a machine can learn to make predictions and detect patterns by analyzing the data, thus providing valuable insights to solve problems without it being explicitly programmed for the purpose (Ratner 2000). The applications of ML in supply chain planning are:

- Carbonneau et al. (2008) used machine learning to incorporate the impact of the "bullwhip effect" on demand forecasting at the end of a supply chain to curb the increasing swings in demand at the level of stockiest retailers' yield supply chain inefficiencies.
- It gives rise to new sources of income funds by executing a self-sufficient supply chain that can consistently monitor excessive lead times for every item depending upon the historical data.
- Increase customer service levels by capturing the precise swings in demand for new items by utilizing algorithms dependent on reaction functions of pre-selling advertisement campaigns to optimize inventory levels and reorder quantities.

2.3.3 Deep Learning

Another offshoot of machine learning, deep learning, leverages inductive learning algorithms and neural networks to train huge unstructured datasets in either a "supervised, semi-supervised or unsupervised" environment, drawing initial inferences and learning from the inaccuracies against actuals then applying that learning to improve the conclusions in the testing dataset. Its applications include:

- Pomerleau (1993) used Artificial Neural Networks as an algorithm utilizing AI techniques to "steer a land vehicle along a single lane on a highway by mimicking the performance of a human driver."
- Artificial Neural Network (ANN) is used to decide the production planning schedule at the manufacturer level and dispatch planning at the retailer level by determining the estimated lot size between successive SC processes (Rohde 2004).

2.3.4 Fuzzy Logic

Fuzzy logic tests the level of "goodness" or "badness" of the decision variable in question by using the exhaustive set of historical data and expert knowledge, leading to the formation of an approximate solution for the variable that handles the ambiguity and uncertainty in the decision-making process (Tanaka 1997). It expresses the decision in a compliment set of binary logic, somewhere between a yes or no, hence named fuzzy or partial. The applications of Fuzzy Logic in supply chain decision-making are as follows:

- The traveling salesman problem (TSP) to follow the shortest route for distribution in the upstream supply chain has been solved with an approximate solution with the help of fuzzy logic since the constraints attached to the problems are uncertain (Michalewicz and Fogel 2000).
- Some other sought-after applications of fuzzy logic in the field of SCM are performance measurement, management, and selection of suppliers (Lau et al. and Wong 2002), controlling the inventory costs (Wang and Shu 2005), and on-time fulfilment of customer orders (Amer et al. 2008).

2.3.5 Agent-based Systems

A robust solution based on AI divides a problem into various subsets of smaller problems. Each sub-problem is addressed with the help of a targeted mix of knowledge, methodology, and resources to arrive at an optimal solution; hence it follows distributed problem-solving technique (Reis 1999). Its applications in supply chain planning are:

- Work allocation issues on the shop floor and planning the production schedule based on the availability of skilled labor have been solved by using a distributed sub-problem methodology (Lima et al. 2011).
- Supply chain relationships, customer relationships, and outsourcing relationship management have been formulated as sub-problems to maintain information flow and curb the uncertainty associated with the supply chain decision-making process (Baxter et al. 2003, Ghiassi and Spera 2003).

2.4 Data Analytics Applications in Supply Chain Processes

2.4.1 Warehouse Management

Proper warehousing is a crucial activity to measure the health of any logistics firm. It requires its managers to have a decent idea of the inventory, the demand, the supply, and the difficulties in maintaining a smooth and effective flow of information across the logistics domain. Machine learning and artificial intelligence prove helpful because the knowledge required for managing the warehouses essentially relies upon the intelligence that can be accumulated from the data (stock count, inventory data, personnel information, labor count, etc.). The data can examine vital information and present valuable insights on some fundamental aspects of warehousing, like overstocking or understocking of inventory for various classes of products, and even the sort of inventory like for raw materials or finished products, including the traits of life span and depreciation of perishable goods when in storage. There are times when the process becomes so simple that AI can assume control over the whole procedure, including the final output. This wipes out the requirement for human mediation. With this disposal of human intervention, the likelihood of a mistake is significantly decreased. One good precedent would be the execution in the Amazon consolidation warehouses with the help of robots that perform the entire range of "pick and drop" operations from specific locations in the warehouses to filling the containers for maximum capacity utilization.

2.4.2 Improving Inventory Planning

One of the most challenging decisions in the lifecycle of a supply chain manager is estimating the amount of inventory to be stocked, considering the demand for the product. The demand is a function of seasonal variability, the price, and the pace at which the product is being consumed in the market (Chen and Simchi-Levi 2004). Predicting how much quantity to order and keep in stock is known as Economic Order Quantity (EOQ) and has been studied in-depth in the literature (Philip 1974, Manna and Chaudhuri 2006) as a make-or-break decision affecting the sales and profitability of a larger scale. AI techniques offer a promising approach for inventory assortment, cutting needless inventory and freeing up space, which reduces the holding cost. Also, the space hence opened can be used for stocking a well-diversified set of products in the warehouse catering to the abrupt demands created in the supply chain. Predictive modeling uses AI to detect consumer tastes and patterns of consumption, which gives an idea of what to stock. Sales opportunities lost due to stock-outs can be recovered by having the right products immediately available; otherwise, it also negatively affects customer loyalty towards the product. Another intriguing application of AI techniques to inventory control and planning was demonstrated by (Teodorovic et al. 2002), who conducted a study on designing fuzzy logic rules to control the dynamic pricing decisions for airlines seats based on the inventory availability, to deliver the special seats at the maximum possible price.

In a nutshell, we can safely assume that a good number of variables can be played with to accurately predict the most optimized inventory level at any point in time. AI is one such solution that can make inventory control planning and processes more efficient.

2.4.3 Network Design and Transportation

For the class of problems, including transportation and network design, literature shows that it is difficult to arrive at a globally optimal solution, leading to it being one of the most popular SC areas for applying AI techniques. Genetic Algorithm usage has been most successful in solving the Traveling Salesman Problem (TSP), which finds the shortest distance to cover all the targets for sale and return to the origin. Similarly, the successfully used cases of the popular meta-heuristic named "ant colony optimization algorithm" on dynamic "vehicle routing" and "minimal spanning tree" problems are also outlined in the literature (Dorigo and Gambardella 1997, Shyu et al. 2003).

Automation seems to be the key to success in every industry, and AI paves the way for automation in the supply chain. For example, AI-based "track and trace" third-party logistics providers are efficiently using tools for performance measurement, on-time delivery, scheduling of forwarding and back-haul services for trucks, and so on. Packaging for transportation is moving towards automation since a variety of logistics robots have been programmed through AI for "pick and pack" service, such that can measure the weight of the package, scan for detecting damaged goods and undertake specific packaging and shipping instructions for a wide range of products.

2.4.4 AI Implications in Purchasing Management

Since the introduction of third-party logistics and outsourcing services in the supply chain domain, a "make or buy" decision has been one of the key performance indicators for a purchase manager. For instance, Niu et al. (2021) analytically model the choice of the restaurant to use the platform logistics service or self-logistics services for online food deliveries. The decision is primarily a function of the market potential measured as weighted profits achieved by producing the goods or services in-house or procuring them from external sources. The decision is also a function of whether the product or the service is the company's core competency or not. Owing to the uncertainty posed by such scenarios, AI has assumed importance in the field of purchase management. One such tool is the expert system that signals the call for "make or buy" by evaluating the end-to-end purchase-level data and presenting insights on crucial factors like suppliers-to-be's performance, transparency in the information exchange with current customers, etc. (Humphreys et al. 2002). Central Order Processing Desk (COPD) is an AI-based concept that helps generate POs for all the locations centrally based on the current inventory count, which includes alerts on low inventory levels, reorder points for raw materials, evaluation of online bids from suppliers, and monitoring the performance of bid-winning suppliers (Cheung et al. 2004).

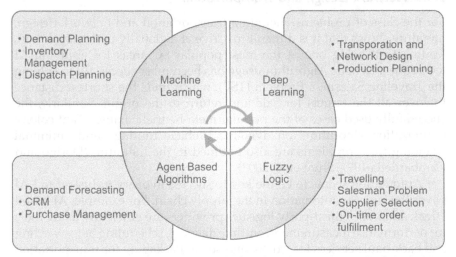

Figure 1: Graphical representation of AI methods and applications in supply chain decision making.

2.4.5 Customer Relationship Management

To increase the loyalty of the customer, it is of prime importance that the customers trust the products and services provided by the company. The valued customers need to be felt special, which can only be established through a long-term and mutually beneficial relationship which is termed Customer Relationship Management which improves service convenience and has a significant impact on a firm's profitability (Mehmood and Najmi 2017).

Baxter et al. (2003) used AI methods and developed an agent-based tool that measured the performance of CRM by weighing its benefits with the cost incurred. The tool used the simulation technique to measure customer experience through social media data and the proportionate profitability achieved for the period. It also expressed the grievances of the customers, which could be improved upon to strengthen the customer relationship.

2.5 Challenges in AI and ML Implementation

The benefits of AI integration in making SCs and their processes highly proactive and automated have triggered its implementation across sectors like food (Li et al. 2010), manufacturing (Kristianto et al. 2017), retail (Cai and Lo 2020), supply procurement (Chopra 2019), healthcare (Nabelsi and Gagnon 2017), oil (Yu et al. 2019), and automotive industry (Werthmann et al. 2017). The conventional techniques find it very complex to make meaning out of large datasets and prescribe analytical observations to optimize the SC processes. Thus, the primary challenge in AI implementation is building the capability of the algorithms to handle large datasets and extract knowledge to enhance the efficiency of the supply chains (Hashem et al. 2015). Big data and AI have a powerful impact on firms and their techniques for managing supply chains. However, Sanders et al. (2019) find that the challenge lies in leveraging AI capabilities to improve transparency and measure the performance of SCs. The effective integration of innovative AI and ML tools in large-scale processes can leverage the data and create a host of social problems for policymakers like the unemployment of unskilled labor since their work is undertaken more effectively by the machines. Ketter et al. (2016) proposed an interdisciplinary approach to deal with these societal problems, training the employees to handle the large datasets and enabling them to take up more sustainable employment opportunities as skilled labor. ML tools can help in the production planning and scheduling given the resource constraints, primarily for the manufacturers engaged in build-to-order and make-to-stock scenarios (Tirkolaee et al. 2021), if the data is maintained with accuracy. They also state that the data on suppliers can also extract crucial information to evaluate their performance and decide on their incentives and future orders. Min (2010) outlines some other challenges of AI applications in SCM—the heavy reliance of AI on machines, which, if programmed incorrectly, can lead to a decision with negative implications, for untrained decision-makers, AI-enabled solutions are also challenging to comprehend and most AI and ML solutions solve narrowly focused SCM problem with a low degree of generalizability. Overall, the challenges associated with AI and ML in SC can be characterized by maintaining complex datasets, policy-related measures, and sustainability concerns in its application across SC domains like inventory management, circular economy, production, supplier selection, transportation, and distribution.

2.6 Conclusion, Limitation, and Future Research

Despite the challenges in AI and ML implementation as discussed in Section 2.5, the field has matured to capture the attention of academics as well as practitioners simultaneously. Today, businesses have started taking significant steps in developing and executing live practical applications of AI in SC and planning to streamline their processes. As an academic discipline, AI in SCM has been projected as one of the fastest-growing areas of interest due to the potential advantages of its implementation, which demands further continued exploration. The emergence of review studies in the recent literature (Tirkolaee et al. 2021, Min 2010, Riahi et al. 2021, Baryannis et al. 2019) reveals many promising avenues for future research. Intelligent agent-based systems can manage the complexities in the real-time pricing of products and services across the SC partners. Game theory can be utilized in agent-based systems to forge strategic SC partnerships ensuring social welfare and profit improvement of all the players. Along similar lines, Riahi et al. (2021) have urged researchers to develop a framework-based approach and map the transformational potential of AI in supply chain decision-making. This framework can identify the strategies the practitioners can follow to assess the bottlenecks in their system that need enhancement, leading to positive SC transformation. Also, emerging technologies like blockchain can remove data-related barriers and improve business processes (Mendling et al. 2018).

While the future with AI seems to be very promising, it is of utmost importance to learn about the limitations of AI before jumping on to its practical usage. A significant amount of investment is required in developing these applications and skill up-gradation since the current skill set of the workforce would prove to be insufficient to comprehend these high-end tools and techniques. To emphasize, integrating AI into the existing systems involves massive effort since human-related activities need to be standardized in a more granular form to achieve positive results. Also, completely depending on computer software may lead to wrong decisions if programmed incorrectly. Artificial intelligence is now discovering its uses even in everyday activities like setting the alarm using personal assistants like "Google Assistant" or "Siri". Despite its presence for a long time, SCM as a field has not exploited the benefits of using AI. The well-defined tactical and operational SC problems have used AI to increase profit. AI has drawn attention from both practitioners and academicians due to the profound benefits it promises. Hence, there are plenty of fertile opportunities that can be explored at the intersection of AI adoption and SC decision-making to address the concerns of SC stakeholders. With this pace of growth, AI seems to have a very positive effect on the SCM area in the future.

References

Amer, Y., L. Luong, S. H. Lee, and M. A. Ashraf. 2008. Optimizing order fulfillment using design for six sigma and fuzzy logic. *International Journal of Management Science and Engineering Management* 3(2): 83–99. https://doi.org/10.1080/17509653.2008.10671038.

Baryannis, G., S. Validi, S. Dani, and G. Antoniou. 2019. Supply chain risk management and artificial intelligence: state of the art and future research directions. *International Journal of Production Research* 57(7): 2179–2202. https://doi.org/10.1080/00207543.2018.1530476.

Baxter, N., D. Collings, and I. Adjali. 2003. Agent-Based Modelling—Intelligent Customer Relationship Management. *BT Technology Journal* 21(2): 126–32. https://doi.org/10.1023/A:1024455405112.

Cai, Y. J., and C.K.Y. Lo. 2020. Omni-channel management in the new retailing era: a systematic review and future research agenda. *International Journal of Production Economics* 229 (November): 107729. https://doi.org/10.1016/J.IJPE.2020.107729.

Carbonneau, R., K. Laframboise, and R. Vahidov. 2008. Application of machine learning techniques for supply chain demand forecasting. *European Journal of Operational Research* 184(3): 1140–54. https://doi.org/10.1016/J.EJOR.2006.12.004.

Čerka, P., J. Grigiene, and G. Sirbikyte. 2015. Liability for damages caused by artificial intelligence. *Computer Law & Security Review* 31(3): 376–89. https://doi.org/10.1016/J.CLSR.2015.03.008.

Chen, X., and D. Simchi-Levi. 2004. Coordinating inventory control and pricing strategies with random demand and fixed ordering cost: the infinite horizon case. *Mathematics of Operations Research* 29(3): 698–723. https://doi.org/10.1287/moor.1040.0093.

Cheung, C. F., W. M. Wang, V. Lo, and W. B. Lee. 2004. An agent-oriented and knowledge-based system for strategic e-procurement. *Expert Systems* 21(1): 11–21. https://doi.org/10.1111/J.1468-0394.2004.00259.X.

Chopra, A. 2019. AI in supply procurement. *Proceedings - 2019 Amity International Conference on Artificial Intelligence, AICAI 2019*, no. February 2019: 308–16. https://doi.org/10.1109/AICAI.2019.8701357.

Dorigo, M., and L. M. Gambardella. 1997. Ant colony system: a cooperative learning approach to the traveling salesman problem. *IEEE Transactions on Evolutionary Computation* 1(1): 53–66. https://doi.org/10.1109/4235.585892.

Ghiassi, M., and C. Spera. 2003. Defining the internet-based supply chain system for mass customized markets. *Computers & Industrial Engineering* 45(1): 17–41. https://doi.org/10.1016/S0360-8352(03)00017-2.

Hashem, I. A. T., Ibrar Yaqoob, N. B. Anuar, S. Mokhtar, A. Gani, and S. U. Khan. 2015. The rise of 'Big Data' on cloud computing: review and open research issues. *Information Systems* 47 (January): 98–115. https://doi.org/10.1016/J.IS.2014.07.006.

Humphreys, P., R. McIvor, and G. Huang. 2002. An expert system for evaluating the make or buy decision. *Computers & Industrial Engineering* 42(2–4): 567–85. https://doi.org/10.1016/S0360-8352(02)00052-9.

Ketter, W., M. Peters, J. Collins, and A. Gupta. 2016. Competitive benchmarking: an is research approach to address wicked problems with big data and analytics. *MIS Quarterly: Management Information Systems* 40(4): 1057–80. https://doi.org/10.25300/MISQ/2016/40.4.12.

Kristianto, Y., A. Gunasekaran, and P. Helo. 2017. Building the 'Triple R' in global manufacturing. *International Journal of Production Economics* 183 (January): 607–19. https://doi.org/10.1016/J.IJPE.2015.12.011.

Lau, H. C. W., W. K. Pang, and C. W. Y. Wong. 2002. Methodology for monitoring supply chain performance: a fuzzy logic approach. *Logistics Information Management* 15(4): 271–80. https://doi.org/10.1108/09576050210436110.

Lezoche, M., H. Panetto, J. Kacprzyk, J. E. Hernandez, and M. M. E. A. Díaz. 2020. Agri-Food 4.0: A survey of the supply chains and technologies for the future agriculture. *Computers in Industry* 117 (May): 103187. https://doi.org/10.1016/J.COMPIND.2020.103187.

Li, Y., M. R. Kramer, A. J. M. Beulens, and J. G. A. J. Van Der Vorst. 2010. A framework for early warning and proactive control systems in food supply chain networks. *Computers in Industry* 61(9): 852–62. https://doi.org/10.1016/J.COMPIND.2010.07.010.

Lima, R. M., R. M. Sousa, and P. J. Martins. 2011. Distributed production planning and control agent-based system. *Https://Doi.Org/10.1080/00207540600788992* 44(18-19): 3693–3709. https://doi.org/10.1080/00207540600788992.

Luger, G. F. 2002. *Artificial intelligence: structures and strategies for complex problem solving*. 4th ed. Essex, England: Pearson Education Limited.

Manna, S. K., and K. S. Chaudhuri. 2006. An EOQ model with ramp type demand rate, time dependent deterioration rate, unit production cost and shortages. *European Journal of Operational Research* 171(2): 557–66. https://doi.org/10.1016/j.ejor.2004.08.041.

Mehmood, S. M. and A. Najmi. 2017. Understanding the impact of service convenience on customer satisfaction in home delivery: evidence from Pakistan. *International Journal of Electronic Customer Relationship Management* 11(1): 23–43. https://doi.org/10.1504/IJECRM.2017.086752.

Mendling, J., I. Weber, W. V. Der Aalst, J. Vom Brocke, C. Cabanillas, F. Daniel, S. Debois, 2018. Blockchains for business process management - challenges and opportunities. *ACM Transactions on Management Information Systems (TMIS)* 9(1). https://doi.org/10.1145/3183367.

Michalewicz, Z. and Fogel, D.B. 2000. *How to solve it: modern heuristics*. Heidelberg, Germany: Springer-Verlag.

Min, H. 2010. Artificial intelligence in supply chain management: theory and applications. *International Journal of Logistics: Research and Applications* 13(1): 13–39. https://doi.org/10.1080/13675560902736537.

Nabelsi, V., and S. Gagnon. 2017. Information technology strategy for a patient-oriented, lean, and agile integration of hospital pharmacy and medical equipment supply chains. *International Journal of Production Research* 55(14): 3929–45. https://doi.org/10.1080/00207543.2016.1218082.

Ni, D., Z. Xiao, and M. K. Lim. 2019. A systematic review of the research trends of machine learning in supply chain management. *International Journal of Machine Learning and Cybernetics* 11(7): 1463–82. https://doi.org/10.1007/S13042-019-01050-0.

Niu, B., Q. Li, Z. Mu, L. Chen, and P. Ji. 2021. Platform logistics or self-logistics? restaurants' cooperation with online food-delivery platform considering profitability and sustainability. *International Journal of Production Economics* 234 (April): 108064. https://doi.org/10.1016/j.ijpe.2021.108064.

Philip, G. C. 1974. A generalized eoq model for items with weibull distribution deterioration. *AIIE Transactions* 6(2): 159–62. https://doi.org/10.1080/05695557408974948.

Pomerleau, D. A. 1993. *Neural Network Perception for Mobile Robot Guidance*. Dordrecht, The Netherlands: Kluwer.

Ratner, B. 2000. A comparison of two popular machine learning methods. *DM STAT-1 – Online Newsletter about Quantitative Methods in Direct Marketing*, 4.

Reis, B. Y. 1999. A multi-agent system for online modeling, parsing and prediction of discrete time series data. *In*: Mohammadian, M. (ed.). *Computational intelligence for modeling, control and automation*. Amsterdam, The Netherlands: IOS Press.

Riahi, Y., T. Saikouk, A. Gunasekaran, and I. Badraoui. 2021. Artificial intelligence applications in supply chain: a descriptive bibliometric analysis and future research directions. *Expert Systems with Applications* 173 (February): 114702. https://doi.org/10.1016/j.eswa.2021.114702.

Rohde, J. 2004. Hierarchical supply chain planning using artificial neural networks to anticipate base-level outcomes. *OR Spectrum* 26(4): 471–92. https://doi.org/10.1007/s00291-004-0170-x.

Russell, S. and Norvig, P. 1995. *Artificial Intelligence: A Modern Approach*. Upper Saddle River, NJ: Prentice-Hall.

Samuel, A.L. 1995. Some studies in machine learning using the game of checkers. *In*: Luger, G. F. (ed.). *Computation and Intelligence: Collected Readings*. Menlo Park, CA: AAAI Press/The MIT Press.

Sanders, N. R., T. Boone, R. Ganeshan, and J. D. Wood. 2019. Sustainable supply chains in the age of AI and digitization: research challenges and opportunities. *Journal of Business Logistics* 40(3): 229–40. https://doi.org/10.1111/JBL.12224.

Shyu, S. J., P. Y. Yin, B. M.T. Lin, and M. Haouari. 2003. Ant-tree: an ant colony optimization approach to the generalized minimum spanning tree problem. *Journal of Experimental and Theoretical Artificial Intelligence* 15(1): 103–12. https://doi.org/10.1080/0952813021000032699.

Singh, S. P., J. Rawat, M. Mittal, I. Kumar, and C. Bhatt. 2022. Application of AI in SCM or supply chain 4.0, 51–66. https://doi.org/10.1007/978-3-030-85383-9_4.

Tanaka, K. 1997. *An introduction to fuzzy logic for practical applications*. New York, NY: Springer-Verlag.

Teodorovic, D., J. Popovic, G. Pavkovic, and S. Kikuchi. 2002. Intelligent airline seat inventory control system. *Transportation Planning and Technology* 25(3): 155–73. https://doi.org/10.1080/0308106022000018991.

Tirkolaee, E. B., S. Sadeghi, F. M. Mooseloo, H. R. Vandchali, and S. Aeini. 2021. Application of machine learning in supply chain management: a comprehensive overview of the main areas. *Mathematical Problems in Engineering* 2021 (Ml). https://doi.org/10.1155/2021/1476043.

Wang, J., and Y. F. Shu. 2005. Fuzzy decision modeling for supply chain management. *Fuzzy Sets and Systems* 150(1): 107–27. https://doi.org/10.1016/J.FSS.2004.07.005.

Werthmann, D., D. Brandwein, C. Ruthenbeck, B. Scholz-Reiter, and M. Freitag. 2017. Towards a standardised information exchange within finished vehicle logistics based on RFID and EPCIS. *International Journal of Production Research* 55(14): 4136–52. https://doi.org/10.1080/00207543.2016.1254354.

Yu, L., Y. Zhao, L. Tang, and Z. Yang. 2019. Online big data-driven oil consumption forecasting with google trends. *International Journal of Forecasting* 35(1): 213–23. https://doi.org/10.1016/J.IJFORECAST.2017.11.005.

CHAPTER 3

Assessing Relations of Lean Manufacturing, Industry 4.0 and Sustainability in the Manufacturing Environment

Vikas Swarnakar

Contents

3.1 Introduction

Nowadays, awareness about sustainability and demand for sustainable products has pressured manufacturing organizations to change their operations. Further, Indian manufacturing organizations are continuously suffering pressure from Indian governments as well as their stakeholders to adopt sustainability in the manufacturing process (Swarnakar et al. 2019a,b). Production philosophies such as LM and Industry 4.0 influence

Industrial and Systems Engineering, Khalifa University, Abu Dhabi, United Arab Emirates.
  Email: vikkiswarnakar@gmail.com

sustainability pillars. "LM is a philosophy that considers the utilization of service and resources for any goal with value creation for the customers" (Sriparavastu and Gupta 1997, Swarnakar et al. 2020a). It is a strategy that eliminates the different wastes from the manufacturing process and enhances the bottom-line results of the industry through its set of well-known tools and techniques (Vinodh and Swarnakar 2015, Shukla et al. 2021). Furthermore, the bottom-line results are improved by systematically streamlining the production line by applying Lean tools and techniques based on human intervention (Antony et al. 2021). A human decision is very important when deciding either at the strategic or operational level.

However, the Lean implemented organization forgets human aspects in the present and mainly focuses on waste elimination (Varela et al. 2019). "The main principle of Industry 4.0 is the integration of both horizontal and vertical production systems driven by real-time data interchangeability and flexible manufacturing systems to enable customized production" (Varela et al. 2019). The data plays several roles such as decision making, prioritization of production orders, task optimization, and maintenance work. The adaptation of Industry 4.0 helps to upgrade technologies through the application of Big Data analysis, high digitalization, automation robots, artificial intelligence, etc. These could be achieved by proper infrastructure such as industrial internet of things (IIoT), Internet of Things (IoT), Artificial intelligence (AI), Cyber-physical systems (CPS), Radio Frequency Identification (RFID), Cloud computing (CC), etc. (Shrouf et al. 2014). Sustainability can be defined as a development that fulfills the present needs without sacrificing future generations (Swarnakar et al. 2019, 2020b). Sustainability is a key driver of innovation. It can be measured by its three main pillars, which are "economic, social, and environmental" (Swarnakar et al. 2021). It is difficult to choose a key measure/Key Performance Index (KPI) to measure the sustainability of the organization. According to Swarnakar et al. 2019, the production's philosophies influence sustainability pillars, but they do not discuss how it can be achieved or how this approach relates to each other. This study addresses these issues and evaluates the relationship between "LM and sustainability" and "Industry 4.0 and sustainability".

3.2 Literature Review

The review of literature encompasses two parts: (a) a review on LM and sustainability; and (b) a review on Industry 4.0 and Sustainability.

3.2.1 Review of LM and Sustainability

This section encapsulates the relationship between LM practices and sustainability dimensions. Swarnakar and Vinodh (2016) used LM to eliminate process waste from manufacturing organizations. Several manufacturing companies have reduced waste in their production processes, increasing their competitiveness (Souza and Alves 2018, Swarnakar 2021). Nevertheless, sustainability in manufacturing organizations is considered a new LM frontier (Martínez-Jurado and Moyano-Fuentes 2014, Swarnakar et al. 2021a). The manufacturing industries that have implemented LM in their processes to improve the bottom-line results also want to see it as socially responsible (Singh et al. 2021). The increase in productivity and savings in total cost is essential for the economic survival of manufacturing organizations (Swarnakar et al. 2021b). However, these tasks should be achieved by mitigating negative social and environmental impacts. Therefore, several authors introduced LM influences towards sustainability. The various influences of LM in the three sustainability perspectives are summarised in Table 1.

Table 1: Influence of LM on Three Sustainability Dimensions.

Dimension	Influence	References
Economic	Decrease operational cost	Azevedo et al. (2012), Gupta et al. (2018)
	Increase profit	Pampanelli et al. (2014)
Social	Increase the quality of work condition	Ioppolo et al. (2014)
	Decrease working accidents	James et al. (2014)
Environmental	Decrease industrial waste	Azevedo et al. (2012), Gupta et al. (2018)
	Decrease non-renewable energy consumption	Ioppolo et al. (2014)

3.2.2 Review of Industry 4.0 and Sustainability

This section outlines the relation between Industry 4.0 practices and sustainability dimensions. Industry 4.0 reflects current trends towards a highly automated data exchange in manufacturing processes and technologies by considering various systems such as CPS, IoT, CC, IIoT, and AI. In the present and upcoming eras, continuous changes are expected in the daily life of humans and industries (Jabbour et al. 2013). The question, in this context, is: can Industry 4.0 help to transfer positive changes and revolutionize the sustainable manufacturing pathway? This study explores the contributions that might arise from numerous authors according to the literature and then attempts to evaluate the relation

Table 2: Influence of Industry 4.0 in Three Sustainability Dimensions.

Dimension	Influence	References
Economic	Decrease operational cost Increase in profit	Shrouf et al. (2014), Waibel et al. (2017) Brettel et al. (2016)
Social	Increase the quality of work condition Decrease working accidents	Kiel et al. (2017) Brettel et al. (2016)
Environmental	Decrease industrial waste Decrease non-renewable energy consumption	Shrouf et al. (2014), Waibel et al. (2017) Fritzsche et al. (2018)

between Industry 4.0 and sustainability. Prevalent authors have revealed the relation between Industry 4.0 and sustainability. Various relations and influences of Industry 4.0 in the three continuums of sustainability perspectives are summarized in Table 2.

3.2.3 Research Gap

The inference observed from the detailed literature review is that many authors used LM constructs, Industry 4.0 constructs, and sustainability indicators and evaluated them in different areas. Based on the author's knowledge, no study has been reported in the literature that evaluates the relation of LM and sustainability and Industry 4.0 and sustainability in the context of Indian automotive component manufacturing organizations. This gap has been formulated to perform this study. Numerous discrepancies have been identified through a review of literature, and several authors utilized LM and Industry 4.0 constructs, and sustainability indicators to evaluate them in various domains. Furthermore, there has been little to no study reported to evaluate the relationship between LM and sustainability and Industry 4.0 and sustainability in the context of Indian manufacturing organizations. Additionally, no thorough and exploratory mixed-method research design has been performed to shed light on the suggested triangulation relation of LM and sustainability and Industry 4.0. These research gaps motivated the authors to conduct the current research.

3.3 Methodology

3.3.1 Framework of the Research

An essential purpose of researchers is to consider the methodological framework and the most appropriate pathway for the analysis.

Furthermore, a straightforward statement of intent for the research is essential. Rowlands (2005) asserts that, in each research work, the first step is to evaluate the research problem, which leads to the selection of a feasible research approach (Swarnakar and Vinodh 2016). We used a mixed-method framework in this analysis, and an exploratory design approach was determined to be the most relevant and successful research method/ approach. Based on the objective of the present research, the elements of LM, Industry 4.0, and Sustainability were incorporated into an "open-ended questionnaire survey", and those key elements were analyzed and finalized throughout the descriptive statistical approach. Generally, the exploratory design approach applies as a path to delineating an occurrence or phenomenon in the trajectory of a causal relationship where survey or empirical validation may be too complex or unpredictable.

3.3.2 Research Method

Saunders (2003) verifies that "different methodologies are extensively used for management analysis and choosing the right one relies on the study's questions and objectives". A questionnaire survey has been used, which was sent via E-Mail. The survey favors large random samples because it has the most consistent results of what is valid in the population (King et al. 2014). Figure 1 depicts the study methods used for this study in the form of a flowchart. The detailed procedure is outlined in the following parts.

3.3.3 Survey Structure

The survey for this study included 27 questions about the respondents' and organizations' contexts, as well as the relationships between LM, Industry 4.0, and the Sustainability approach. The initial phase of the research was collecting data on the respondent's experience and qualifications, role in the business of the organization, and the extent to which LM, Industry 4.0, and Sustainability approaches had been used in the organization. The second portion of the questionnaire examined the key metrics used throughout the company and asked the respondent to examine the effectiveness of the program so far to determine whether LM influenced Industry 4.0 and the Sustainability pathway. In this context, a closed-ended questionnaire format has also been used to allow for statistical analysis. In the final section of the questionnaire, the 27 elements of LM, 19 elements of Industry 4.0, and 14 elements of Sustainability from the literature review were characterized on a "5-point Likert scale". Consequently, the replies to the Likert items were typically symmetrical in the form of strongly agreeing (scale 5) to strongly disagreeing (scale 1),

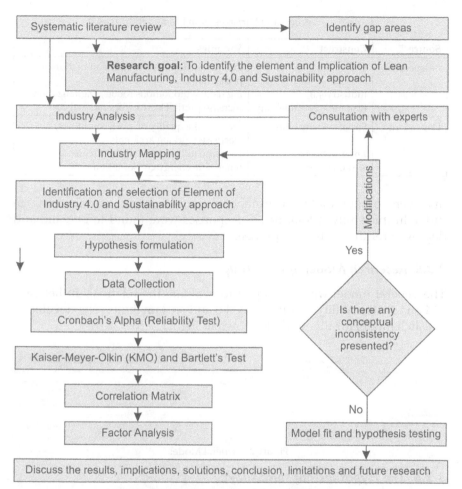

Figure 1: Research methodology flowchart.

embraced with a neutral response as "neither agree nor disagree" (scale 4). The Likert scale value could be interpreted as the ordinal variable.

3.3.4 *Sample and Data Collection Strategy*

In the present study, a set of variables was selected from the literature review for both independent and dependent constructs of LM and sustainability and is presented in Table 3. Next, the questionnaire was designed based on these constructs and variables with the help of three industry experts. The experts have a minimum of 10 years of industrial experience. The questionnaire was designed on a Likert scale of 1 to 5 with the convenience sampling method and then mailed to 199 general

Table 3: Constructs and Variables from LM and Industry 4.0.

Source	Construct	Variables
Independent	LM	Pull production, Waste elimination, Pursue perfection
	Industry 4.0	Big data, Autonomous robots, Digitalization
	Economic sustainability	Increase profit, Decrease operational cost
Dependent	Social sustainability	Increase the quality of work conditions, Decrease working accidents
	Environmental sustainability	Decrease industrial waste, Decrease non-renewable energy consumption

managers of automotive component manufacturing organizations in India. In this study, a total of 137 responses were found to be valid and deployed for the evaluation process.

3.3.5 Research Model of this Study

The general model was developed to evaluate the relationship between LM and sustainability and Industry 4.0 and sustainability. The developed research model for this study is shown in Fig. 2.

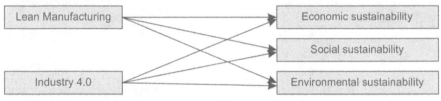

Figure 2: General Model.

3.3.6 The Hypothesis of this Study

The formulation of present hypotheses is prevalently based on reflections on existing business and iterative panel organizations. To further adhere to the experts' assessments, additional evidence from scientific literature was embraced to form and derive the hypotheses. To develop the research hypothesis, the implications of an LM on Industry 4.0 and Sustainability approach deployment outcomes were studied and validated based on the structured review. Based on the general model the following research hypothesis has been formulated and the following hypothesis has been developed.

Hypothesis 1 – H1: The adaptation of the LM approach in automotive industries is positively related to Economic sustainability.

Hypothesis 2 – H2: The adaptation of the LM approach in the automotive industry is positively related to social sustainability.

Hypothesis 3 – H3: The adaptation of the LM approach in the automotive industry is positively related to environmental sustainability.

Hypothesis 4 – H4: The adaptation of the industry 4.0 approach in the automotive industry is positively related to Economic sustainability.

Hypothesis 5 – H5: The adaptation of the industry 4.0 approach in the automotive industry is positively related to social sustainability.

Hypothesis 6 – H6: The adaptation of the industry 4.0 approach in the automotive industry is positively related to environmental sustainability.

3.3.7 Measure of the Study

The questionnaire consisted of an array of questions related to LM and the Industry 4.0 construct, which were prepared and designed by the authors based on various literature reviews. The questionnaire was developed in three sections. The initial phase of the research was collecting data on the respondent's experience and qualifications, role in and business of the organization, and the extent to which LM and Industry 4.0 had been used in the organization. The second section covered questions related to both the LM and Industry 4.0 constructs, and the third phase investigated the key metrics of the sustainability variables. The recipients were asked to rate those questions according to a 5-point Likert scale". The relationships between the two-production philosophy (LM and industry 4.0) and the three sustainability dimensions and their key elements were examined using the developed general model and a thorough review of the literature. Three questions were formed for each production philosophy to measure their importance, i.e., "LM" was comprised of three questions, and a similar rule was applied to Industry 4.0. In addition, for the sustainability pillars, two questions were formed for each dimension. All of these questions are revealed in the results and discussion section.

3.4 Results

Analysis of Survey Outcomes

The collected data and analysis were performed by utilizing "IBM SPSS Statistics 20 Software". The software originally stood for "Statistical Package for the Social Sciences (SPSS)". It is statistical software developed by IBM for advanced analytics, data management, business intelligence, multivariate analysis, and criminal investigation. In the present study, a reliability test was carried out to examine the survey validity. The "Cronbach's Alpha reliability test (Cronbach's Alpha)" analyzes and is an "indicator of reliability, internal consistency reliability, convergent validity, and discriminant validity" (Silva et al. 2014). The reliability test evaluates the characteristics of measurement scales and the elements that make up those scales. An alpha coefficient of 0.6 or higher is perceived to be an adequate level of internal consistency. The analysis revealed an overall Cronbach's alpha coefficient of 0.751 for the LM survey, and 0.748 for Industry 4.0, indicating that the data obtained was appropriate for the research. Additionally, the "Kaiser-Meyer-Olkin (KMO) and Bartlett's tests" were used to determine the appropriateness of performing factor analysis. The analysis shows that the KMO is 0.788 (see Table 4).

The factor analysis was performed to find out the factor loading of both internal and external constructs. Furthermore, multiple regression analysis was performed to evaluate the impact of independent constructs (LM, Industry 4.0) on the dependent variables (Economic, Social and Environmental sustainability). The detailed assessment is revealed in the next section.

Table 4: KMO and Bartlett Test Analysis Results for Independent Variables.

KMO and Bartlett Test		
KMO Measure of Sampling Adequacy		.788
	Approximate Chi-Square	2550.521
Bartlett Test of Sphericity	Degree of freedom	351
	Significance.	.000

Outcomes

Step 1. Factor analysis:

Factor analysis is a technique that is used to reduce many constructs into a fewer number of factors. In this technique, the common variances are extracted from all constructs and put into a common score based on the analysis of the respondent's data. This score is known as factor loading, and it is used for the further analysis process. As per the thumb rule, the value of 0.7 or higher of factor loading reveals that the factor extracted would have been high loading factors, which are perceived to be a significant variable. The evaluation results for independent and dependent variables are displayed in Table 4, Table 5, Table 6, and Table 7.

The calculations of KMO and Bartlett's assessments have been revealed in Table 4. The outcome has indicated that the KMO value is 0.788, which is > 0.7 (acceptable value). Therefore, this data is acceptable for further investigation. Higher levels typically suggest that factor analysis can be relevant when considering and detecting data structure.

Furthermore, the significance level of Bartlett's test has been calculated to be 0.000, which is 0.05 (acceptable limit as per the thumb rule). The result asserts that the data is appropriate for structure detection or measuring sampling adequacy for factor analysis. Consequently, the factor analysis was conducted on IBM SPSS Statistics 20 software, and the analysis result is displayed in Table 5.

The result of factor analysis as independent variables (Table 5) shows that all Cronbach values for independent variables LM and Industry 4.0 are > 0.7, which indicates the internal consistency and reliability of the data, and the data obtained was appropriate for the research. The calculated variance is 69.24%, which is explained as the factor analysis result being good for validation purposes.

The results of the KMO and Bartletts tests are presented in Table 6. The result showed that the KMO value is 0.726, which is > 0.7 (acceptable value). Thus, this data could be perceived to be acceptable for further investigation. The result shows that the data is appropriate for factor analysis. Subsequently, factor analysis was performed, and the evaluation result is presented in Table 7.

The results of factor analysis for dependent variables (Table 7) show that all Cronbach values of DVs are greater than 0.7, confirming that the data are reliable and that there is consistency among the variables. The calculated variance is 73.41%, which is explained as the factor analysis result being good for validation purposes.

Table 5: Factor Analysis Results for the Independent Variable.

S. No.	Independent construct	Factor loading value	Variance %	Cronbach value α
LM			38.03%	0.751
1	Bring Pull production process into an organization	0.927		
2	Elimination of waste using different tools and techniques	0.811		
3	Pursue perfection through continuous process improvement	0.728		
Industry 4.0			31.21%	0.748
1	The big data approach provides different ways to analyze a complex data set	0.812		
2	Autonomous robots perform tasks to a higher degree of anatomy	0.710		
3	Ensure digitalization in the process	0.927		

Table 6: KMO and Bartlett Test Analysis Results for the Dependent Variable.

KMO and Bartlett Test		
Kaiser Meyer Olkin Measure of Sampling Adequacy		.726
Bartlett Test of Sphericity	Approximate Chi-Square	2837.317
	Degree of freedom	379
	Significance.	.000

Table 7: Findings of Factor Analysis for the Dependent Variable.

S. No.	Dependent construct	Factor loading value	Variance %	Cronbach value α
Economic sustainability			25.23%	0.738
1	Increase the profit of the organization	0.913		
2	Decrease operational cost of the organization	0.823		
Social sustainability			23.11%	0.765
1	Increase the quality of work conditions of the organization	0.789		
2	Decrease working accidents in the organization	0.718		
Environmental sustainability			25.07%	0.712
1	Decrease industrial waste in an organization	0.928		
2	Decrease non-renewable energy consumption in the organization	0.831		

Step 2. Multiple regression analysis

In descriptive statistical modeling, the purpose of multiple regression is to understand more about the interrelation between many independent and dependent constructs. In this study, multiple regression analysis was performed using "IBM SPSS Statistics 20" software. The analysis results of the relationship between LM and sustainability and between Industry 4.0 and sustainability are revealed in Table 8.

According to the results of Table 8, the value of R square for economic, social, and environmental sustainability is 49.7%, 22.2%, and 28.1%, respectively, indicating that the implementation of LM and Industry 4.0 in Indian manufacturing organizations explains 49.7% of economic sustainability, 22.2% of social sustainability, and 28.1% of environmental sustainability. Thus, it could be inferred that both production philosophies' executions are primarily weighted and prefer economic sustainability over other sustainability indicators.

Table 8: Findings from Multiple Regression Analysis.

Independent construct	Dependent construct					
	Economic sustainability		Social sustainability		Environmental sustainability	
	Std. coeff. β	Sig.	Std. coeff. β	Sig.	Std. coeff. β	Sig.
Constant		0.869		0.721		0.937
LM	0.061*	0.092	0.079*	0.073	0.062*	0.081
Industry 4.0	0.054*	0.086	0.157	0.211	0.047*	0.093
R-value	0.495		0.397		0.491	
R square value	0.245		0.157		0.079	
F value	2.147		1.325		1.721	
*p < 0.10, Std. Coeff. = Standardized coefficient, Sig. = Significance						

The purpose of the prior hypothesis formulation was to evaluate the relation between LM and sustainability, and between Industry 4.0 and sustainability based on their measurement variables. The hypothesis testing was performed using SPSS 20 incorporated with the multiple regression analysis and the results are revealed as follows:

H1 is accepted since *p < 0.10 and is at the significance level.

H2 is accepted since *p < 0.10 and at the significance level.

H3 is accepted since *p < 0.10 and at the significance level.

H4 is accepted since *p < 0.10 and is at the significance level.

H5 is rejected since p > 0.10 and it is not at the significance level.

H6 is accepted since *p < 0.10 and is at the significance level.

3.5 Conclusion

The study discussed the evaluation of the relationship between LM and sustainability indicators and between Industry 4.0 and sustainability indicators. The evaluation has been performed based on the data collected from various automotive component manufacturing organizations located in India. However, the study utilized factor analysis and multiple regression analysis to assess the interrelations between these approaches. The outcome of this study showed that LM has a positive relationship with all three sustainability indicators, such as economic, social, and environmental. Further, the hypothesis evaluation result indicates that Industry 4.0 has a positive relationship with economic and environmental sustainability and a negative relationship with social sustainability.

The reason behind this negative relationship is that the industry 4.0 approach brings the use of a computer interaction environment and makes digitalization in the process through artificial intelligence and the use of a robot, which consequently reduces the human requirement in manufacturing organizations. Therefore, the outcome indicated a negative relationship between Industry 4.0 and social sustainability. Furthermore, the findings of this study can be summarized as the adaptation of LM to the organization's having positive relationships with all three sustainability indicators. However, it has a greater effect on economic sustainability than environmental sustainability and lastly, social sustainability. Nevertheless, the adaptation of the industry 4.0 approach also has a significant positive relationship with economic and environmental sustainability.

3.6 Implications, Limitations, and Future Work

This study can be an incentive to Indian automotive component manufacturing organizations that struggling to seek and adopt which kinds of continuous improvement approaches are the most appropriate. The results of this study can help decision-makers, stakeholders, and practitioners to have a deeper insight into the relationship between LM and Industry 4.0 in the context of sustainability and have guided to incorporation and adaptation of these continuous improvement frameworks. The limitation of the study could be perceived as obtained results were based on a questionnaire gathered from general managers with small data set. In the future longitude studies with large data sets at various levels of employees of the organization, including other countries could be performed. In addition, the proposed model can be validated by the other MCDM tools such as SEM or DEA technique.

References

Antony, J., V. Swarnakar, E. Cudney, and M. Pepper. 2021. A meta-analytic investigation of lean practices and their impact on organisational performance. *Total Quality Management & Business Excellence* (2021): 1–27. https://doi.org/10.1080/14783363.2021.2003194.

Azevedo, S. G., H. Carvalho, S. Duarte, and V. Cruz-Machado. 2012. Influence of green and lean upstream supply chain management practices on business sustainability. *IEEE Transactions on Engineering Management* 59(4): 753–765.

Brettel, M., M. Klein, and N. Friederichsen. 2016. The relevance of manufacturing flexibility in the context of Industrie 4.0. *Procedia Cirp* 41: 105–110.

Fritzsche, K., S. Niehoff, and G. Beier. 2018. Industry 4.0 and climate change—Exploring the science-policy gap. Sustainability 10(12): 4511.

Gupta, V., G. Narayanamurthy, and P. Acharya. 2018. Can lean lead to green? Assessment of radial tyre manufacturing processes using system dynamics modelling. *Computers & Operations Research* 89: 284–306.

Ioppolo, G., S. Cucurachi, R. Salomone, G. Saija, and L. Ciraolo. 2014c. Industrial ecology and environmental lean management: Lights and shadows. *Sustainability* 6(9): 6362–6376.

Jabbour, C. J. C., A. B. Lopes de Sousa Jabbour, K. Govindan, A. A. Teixeira, and W. R. de Souza Freitas. 2013. Environmental management and operational performance in automotive companies in Brazil: the role of human resource management and lean manufacturing. *Journal of Cleaner Production,* 47: 129–140.

James, J., L. H. Ikuma, I. Nahmens, and F. Aghazadeh. 2014. The impact of Kaizen on safety in modular home manufacturing. The *International Journal of Advanced Manufacturing Technology* 70(1-4): 725–734.

Kiel, D., J. M. Müller, C. Arnold, and K.-I. Voigt. 2020. Sustainable industrial value creation: Benefits and challenges of industry 4.0. *In Digital Disruptive Innovation,* 21(08): 231–270.

King, D. B., N. O'Rourke, and A. DeLongis. 2014. Social media recruitment and online data collection: A beginner's guide and best practices for accessing low-prevalence and hard-to-reach populations. *Canadian Psychology/Psychologie canadienne* 55(4): 240.

Martínez-Jurado, P. J., and J. Moyano-Fuentes. 2014. Lean management, supply chain management and sustainability: a literature review. *Journal of Cleaner Production* 85: 134–150.

Pampanelli, A. B., P. Found, and A. M. Bernardes. 2014. A Lean & Green Model for a production cell. *Journal of Cleaner Production* 85: 19–30.

Rowlands, B. H. 2005. Grounded in practice: Using interpretive research to build theory. *The Electronic Journal of Business Research Methodology* 3(1): 81–92.

Saunders, Peter. Social theory and the urban question. *Routledge,* 2003.

Shrouf, F., J. Ordieres, and G. Miragliotta. 2014. Smart factories in Industry 4.0: A review of the concept and of energy management approached in production based on the Internet of Things paradigm. In 2014 IEEE International Conference on Industrial Engineering and Engineering Management, pp. 697–701. IEEE.

Shukla, V., V. Swarnakar, and A. R. Singh. 2021. Prioritization of Lean Six Sigma project selection criteria using Best Worst Method. *Materials Today: Proceedings* 47(17): 5749–5754.

Silva, G. M., P. J. Gomes, L. F. Lages, and Z. L. Pereira. 2014. The role of TQM in strategic product innovation: an empirical assessment. *International Journal of Operations & Production Management.*

Singh, K., V. Swarnakar, and A. R. Singh. 2021. Lean Six Sigma project selection using Best Worst Method. *Materials Today: Proceedings* 47(17): 5766–5770.

Souza, J. P. Estevam, and J. M. Alves. 2018. Lean-integrated management system: A model for sustainability improvement. *Journal of Cleaner Production* 172: 2667–2682.

Sriparavastu, L., and T. Gupta. 1997. An empirical study of just-in-time and total quality management principles implementation in manufacturing firms in the USA. *International Journal of Operations & Production Management* 17(12): 1215–1232.

Swarnakar, V., and S. Vinodh. 2016. Deploying Lean Six Sigma framework in an automotive component manufacturing organization. *International Journal of Lean Six Sigma* 7(3): 267–293.

Swarnakar, V., A. R. Singh, and A. K. Tiwari. 2019. Assessment of manufacturing process through lean manufacturing and sustainability indicators: Case. *Emerging Trends in Mechanical Engineering: Select Proceedings of ICETME 2018:* 253.

Swarnakar, V., A. R. Singh, and A. K. Tiwari. 2019a. Evaluating importance of critical success factors in successful implementation of Lean Six Sigma framework. *In AIP Conference Proceedings* 2148(1): 030048. AIP Publishing LLC.

Swarnakar, V., S. Vaidya, A. K. Tiwari, and A. R. Singh. 2019b. Assessing critical failure factors for implementing lean six sigma framework in Indian manufacturing organizations. In 3rd IEOM European Conference on Industrial Engineering and Operations Management, 2161–2172.

Swarnakar, V., A. R. Singh, J. Antony, A. K. Tiwari, E. Cudney, and S. Furterer. 2020a. A multiple integrated approach for modelling critical success factors in sustainable LSS implementation. *Computers & Industrial Engineering* 150: 106865.

Swarnakar, V., A. K. Tiwari, and A. R. Singh. 2020b. Evaluating critical failure factors for implementing sustainable lean six sigma framework in manufacturing organization: A case experience. *International Journal of Lean Six Sigma* 11(6): 1069–1104.

Swarnakar, V. 2021. Assessment of critical failure factors for implementing lean six sigma in manufacturing industry: A case study. *International Journal of Industrial Engineering* 3(1).

Swarnakar, V., A. Bagherian, and A. R. Singh. 2021a. Modeling critical success factors for sustainable LSS implementation in hospitals: an empirical study. *International Journal of Quality & Reliability Management* (2021a). https://doi.org/10.1108/IJQRM-04-2021-0099

Swarnakar, V., A. R. Singh, and A. K. Tiwari. 2021b. Evaluation of key performance indicators for sustainability assessment in automotive component manufacturing organization. *Materials Today: Proceedings* 47(17): 5755–5759.

Swarnakar, V., A. R. Singh, J. Antony, A. K. Tiwari, and E. Cudney. 2021c. Development of a conceptual method for sustainability assessment in manufacturing. *Computers & Industrial Engineering* 158: 107403.

Varela, L., A. Araújo, P. Ávila, H.Castro, and G. Putnik. 2019. Evaluation of the relation between lean manufacturing, Industry 4.0, and sustainability. *Sustainability* 11(5): 1439.

Vinodh, S., and Vikas Swarnakar. 2015. Lean Six Sigma project selection using a hybrid approach based on fuzzy DEMATEL–ANP–TOPSIS. *International Journal of Lean Six Sigma* 6(4): 313–338.

Waibel, M. W., L. P. Steenkamp, N. Moloko, and G. A. Oosthuizen. 2017. Investigating the effects of smart production systems on sustainability elements. *Procedia Manufacturing* 8: 731–737.

CHAPTER 4

Role of Artificial Intelligence in Supply Chain Management

Pratibha Garg, Neha Gupta* and Mohini Agarwal

Contents

4.1 Introduction

AI consists of a set of computational technologies developed to sense, learn, reason, and act appropriately. With the technological advancement in mobile computing, the capacity to store huge data on the internet, cloud-based machine learning and information processing algorithms, etc. AI has been integrated into many sectors of business and has been proven to reduce costs, increase revenue, and enhance asset utilization. Artificial intelligence (AI) refers to the simulation of human intelligence

Amity School of Business, Amity University, Uttar Pradesh, Noida.
* Corresponding author: ngngupta4@gmail.com

in machines that are programmed to think like humans and mimic their actions. The term may also be applied to any machine that exhibits traits associated with a human mind such as learning and problem-solving. The ideal characteristic of artificial intelligence is its ability to rationalize and take actions that have the best chance of achieving a specific goal. A subset of artificial intelligence is machine learning, which refers to the concept that computer programs can automatically learn from and adapt to new data without being assisted by humans. Deep learning techniques enable this automatic learning through the absorption of huge amounts of unstructured data such as text, images, or video.

A supply chain is a network between a company and its suppliers to produce and distribute a specific product to the final buyer. This network includes different activities, people, entities, information, and resources. The supply chain also represents the steps it takes to get the product or service from its original state to the customer. Companies develop supply chains so they can reduce their costs and remain competitive in the business landscape. Supply chain management is a crucial process because an optimized supply chain results in lower costs and a faster production cycle. The evolution and increased efficiencies of supply chains have played a significant role in curbing inflation. As efficiencies in moving products from A to B increase, the costs in doing so decrease, which lowers the final cost to the consumer. While deflation is often regarded as a negative, supply chain efficiencies are one of the few examples where deflation is a good thing. As globalization continues, supply chain efficiencies become more optimized, which keeps the pressure on input prices.

4.2 Applications and Challenges in the Adoption of Artificial Intelligence

The applications for artificial intelligence are endless. The technology can be applied to many different sectors and industries. AI is being tested and used in the healthcare industry for dosing drugs and different treatments in patients, and for surgical procedures in the operating room. Other examples of machines with artificial intelligence include computers that play chess and self-driving cars. Each of these machines must weigh the consequences of any action they take, as each action will impact the result. In chess, the result is winning the game. For self-driving cars, the computer system must account for all external data and compute it to act in a way that prevents a collision. Artificial intelligence also has applications in the financial industry, where it is used to detect and flag activity in banking and finance such as unusual debit card usage and large account deposits— all of which help a bank's fraud department. Applications for AI are also being used to help streamline and make trading easier. This is done by making the supply, demand, and pricing of securities easier to estimate. AI enables retail and manufacturing businesses in making smarter decisions, with more accurate and real-time forecasting, to improve supply management, define impactful thematic promotions, optimizing assortment and pricing. AI also makes operations more efficient, because of robotics and process optimization which enhances productivity and reduces manual labor costs. The use of interactive Robots in the warehouse and store are well known. The advancement in enhanced vision is enabled by more powerful computers, new algorithmic models, and large training data sets. Within the field of computer vision, object recognition and semantic segmentation—that is, the ability to categorize object type, such as distinguishing a tool from a component, and have recently advanced significantly in their performance (Wen et al. 2018, Bughin et al. 2017). They allow robots to behave appropriately for the context in which they operate, for example by recognizing the properties of the materials and objects they interact with. They are flexible and autonomous and capable of safely interacting with the real world and humans. Companies like Swisslog, DHL, etc., are using these technologies efficiently (Bughin et al. 2017).

However, there are many hurdles yet to be crossed before we get the full potential of AI. The first and foremost is to gain the confidence of the stakeholders, i.e., managers and employees, and those involved in the regulatory and policy-making boards. Robots are gradually being adapted to perform packaging and delivery. However, we still do not know how to address technical difficulties.

4.3 Literature Review

Supply chain management (SCM) is one of the most challenging fields which emphasizes interactions among different sectors, primarily marketing, logistics, and production. Therefore, success in SCM lies in the overall success of any business. However, with the consistent changes in business practices like lean management and just-in-time philosophy both in production and logistics, globalization, adverse events, i.e., frequent natural disasters, political instability, etc. SCM always needs to develop an adequate solution to mitigate such challenges. In recent years technologies like Artificial Intelligence (AI) have proved immensely valuable to SCM. Today, globalization and technological innovations call for improved organizational adaptability and more flexible and advanced systems relative to manufacturing, logistics, engineering, information, and process technology (Momme 2002). To increase their performance, firms are outsourcing their non-core competencies activities and are focusing on their own business. The technological advancement in mobile computing, artificial neural networks, robotics, storage of huge data on the internet, cloud-based machine learning, information processing algorithms, etc., has propelled the use of AI in various business sectors (Dash et al. 2019). Dominguez et al. (2010) found that the alignments in the supply chain are complex as they include a high number of partners. This is because the supply chain includes not only the horizontal one of the core products (jewels) but also the vertical supply chain of RFID tags. RFID projects need a matrix supply chain, both horizontal and vertical, to manage a product composed of a jewel plus a tag. These alignments are incremental and are the results of the interactions between different types of alignment in the supply chain. Each intra-organizational alignment has an impact on the other intra-organizational alignments of the partners in the same supply chain. Finally, the extended inter-organizational alignment model can be compared to an eco-system with continuous micro intra-organizational alignments that lead to a global and incremental inter-organizational alignment.

Kannabiran and Bhaumik (2005) discussed the role of IT in their Supply Chain, strategic outsourcing, etc. They concluded with the fact that SCM strategies will be successful through strong top management support to initiate key changes and provide necessary resources. Further, cross-functional teams are effective in implementing and managing the change in key SCM projects with specific objectives. The shared responsibility of business partners plays a major role in achieving a high degree of effectiveness and efficiency in the entire supply chain.

Correa (2020) found that for a robust supply chain, most companies intend to adopt technologies like AI, the Internet of Things, big data, Cloud, crowdsourcing, etc.

Mittal et al. (2020) cater to the pressing need of manufacturing SMEs for targeted support by proposing a novel SM paradigm adoption framework. Adopting the SM paradigm provides SMEs with a competitive edge and is a necessary step to be considered as collaborators in digital supply networks. Raman et al. (2018) have focused on the impact of big data on the supply chain industry in terms of its potential to create new value by enhancing operational excellence, enabling cost-saving measures, increasing customer satisfaction and real-time visibility, and narrowing the gap between supply and demand chain management, thereby influencing the adoption of big data technology. However, the fact that big data technology is in its nascent stage, together with the initial monetary costs and the lack of knowledge concerning its implementation, hampers its adoption in the supply chain industry. The researcher has further reviewed various factors to analyze the relation of AI with the supply chain

4.3.1 Forecast Demand

AI has been effectively used in projection and forecasting. Organizations are always keen to balance both supply and demand. Therefore, a better forecast is needed for its supply chain and manufacturing. As AI can process, analyze (automatically) and more importantly, predict data, it provides an accurate and reliable forecasting demand, which allows businesses to optimize their sourcing in terms of purchases and order processing, therefore, reducing costs related to transportation, warehousing, and supply chain administration, etc. In addition, it discerns trends and patterns which help to design better retailing and manufacturing strategies. As these demand forecasts are so accurate, they do not lose a sale because of product unavailability. Machine learning approaches not only incorporate historical sales data and the setup of the supply chains but also rely on near-real-time data regarding variables such as advertising campaigns, prices, and local weather forecasts (Bughin et al. 2017). Otto a German online retailer manages to reduce 90% of its inventory using such an application. The AI forecasts are so reliable that Otto builds its inventory in anticipation of the orders, more interestingly relying on AI without any human intervention (Burgess 2018). AI has also been used in R&D departments, to quickly assess whether a prototype would be likely to succeed or fail in the market and if so why. More importantly, it delivers more efficient designs by eliminating waste in the design process. By doing so AI has played an important role in smart manufacturing (Kusiak 2018).

4.3.2 Supplier Relations

A supply chain is a network between a company and its suppliers to produce and distribute a specific product to the final buyer. This network includes different activities, people, entities, information, and resources. The supply chain also represents the steps it takes to get the product or service from its original state to the customer. Companies develop supply chains so they can reduce their costs and remain competitive in the business landscape (Van Hoek 1998, Chen and Paulraj 2004). Effective supply chain management is dependent on many internal and external environmental variables of an organisation. Uncertainty in demand, supply and technology are one of the fundamental issues to be managed (Van Hoek 1998, Chen and Paulraj 2004) to arrive at a responsive supply chain. Understanding changing customer needs and accordingly designing supply chains to deliver products and services will help organisations to outperform their competition (Carson et al. 1998, Sinha and Subash Babu 1998, Tan et al. 1999). The role of top management in understanding the complexities and willingness to support changes in the existing supply chain is a critical requirement (Monaszka et al. 1993, Krause 1999).

Moreover, inter-organisational communication (Krause 1999, Carr and Smeltzer 1999), the use of cross-functional teams in managing relationships (Krause and Ellram 1997, Ellinger 2000) and a high level of involvement of suppliers (Swink 1999, Shin et al. 2000) are some of the determinants of effective SCM. Many researchers have emphasised the need for coordination amongst interdependent units of different organisational authority and responsibility in a supply chain. SCM involves varied levels of coordination of activities within and between organisations in the supply chain (Cooper et al. 1997, Simatupang and Sidharan 2002). A recent study of supply chain coordination in a fashion firm has concluded that inter-functional conflicts arise because of the differences in expectations of the work domain and ambiguity of decision-making authority. Further, in the era of outsourcing and preparing organisations for the ever-increasing competitive markets, supply chain coordination is a critical capability for organisations (Lee 2002). Thus, researchers have proposed frameworks and approaches to manage the coordination in supply chains (Lee 2000, Hines et al. 2000, Simatupang and Sidharan 2002). The role of AI in seamlessly integrating the supply chain has been emphasised by researchers (Min and Galle 1999, Lee and Whang 2000). Enabling the supply chain using AI as a strategic and capital-intensive initiative is affected by mutual trust and long-term relationships (Laseter 1997, Kilpatrick and Factor 2000, Agarwal and Shankar 2003).

4.3.3 Smart Manufacturing

The use of AI has transformed the manufacturing sector, from virtual assistants to advanced robotics, and has enabled manufacturing companies to produce more with fewer errors to adept demand. Using AI helped them in rapid growth as they can shorten development cycles, improve engineering efficiency, prevent faults, increase safety by automating risky activities, reduce inventory costs with better supply and demand planning, and increase revenue with better sales lead identification and price optimization, etc. (Patel et al. 2018, Bughin et al. 2017). The new concept, i.e., "Intelligent manufacturing" is a smart approach for production where machines are linked to humans, i.e., both machines and humans are working side-by-side with minimal guidance. The best example of intelligent manufacturing is the manufacturing sector of Siemens. The employee manages and controls the production of programmable logic circuits through a virtual factory that replicates the factory floor. Via barcodes, products communicate with the machines that make them, and the machines communicate among themselves to replenish parts and identify problems (Bughin et al. 2017). As high as 75% of the production process is fully automated, and 99.99988% of the logic circuits are defect-free. Similarly, AI and 3D printing have revolutionized customization in manufacturing. Intel has developed Predictive analytics using machine learning a powerful tool to reduce the time required to solve design problems for semiconductor manufacturers (Burgess 2018). Motivo, an artificial intelligence start-up, managed to compress semiconductor design processes from years to a few weeks, saving chip makers the cost of iterations and testing. Using machine learning, aerospace manufacturing industries have developed productivity tools for engineering teams, i.e., team travel norms, team composition, supplier communication, etc. Machine learning has reduced its development costs by unleashing the speed, accuracy, and relevance of products. AI has allowed manufacturers to integrate production and client feedback in real-time to refine product design. With suppliers, AI-based tools provide better accountability throughout the supply chain, which helps aerospace manufacturers as well. For example, manufacturing a jet requires 1000s of parts and procurement of these parts from around the world is a complex challenge. AI technologies like Virtual reality in manufacturing link thousands of different parts most importantly, it provides transparency on supplier machine availability, performance, downtime, etc. It helps in balancing the supply chain and optimising inventories in real-time (Kraus et al. 2018). AI has helped in cutting costs also to a great extent. AI has empowered various business processes reducing human error and hence saving a lot of costs which is sometimes too high in the field of Jewellery. Using AI manufacturers are optimizing the key performance indicators and reviewing them in real-time. For example, tailoring a model using virtual programs help to better predict, identify, and prevent material and staffing bottlenecks and optimize energy consumption. More importantly, it alerts the engineers before problems arise and recommend solutions. AI-enabled manufacturing not only efficiently assembly line practices but also cuts costs, reduces waste, and speeds time to market (O'Reilly et al. 2019). Using machine learning algorithms, collaborative robots, and self-driving vehicles have been proven to improve warehouse costs and reduce inventory levels as well. The best example is when General Electrics turned to Kaggle, the platform for predictive modelling and analytics competitions, and invited data scientists to design new routing and machine learning algorithms for flight planning that optimized fuel consumption by looking at variables such as weather patterns, wind, and airspace restraints. The winning routing algorithm showed a 12% improvement in efficiency over actual flight data (Henriques et al. 2018). The key feature of AI-enabled manufacturing is collaborative agility, which is the ability to adapt almost instantly to changes in demand and the evolution of regulation, input prices, technologies, and other parts of the industry landscape. Now smart Robots are working with humans collaboratively for the mass production of products which is essential for customer-centric products. AI has helped manufacturing plants around the globe in supply chains, and value chains which are more interconnected and collaborative (Klumpp 2018). AI is also not far behind in agriculture. The concept of "e-plants in a box" is a reality which is great for small-scale, low-capital-expenditure, mobile plants that can produce a limited range of products at a competitive cost. Huxley combines machine learning, computer vision, and an augmented reality interface to essentially allow anyone to be a master farmer. More importantly, these e-plants can be transported to markets where demand is temporarily strong and in remote markets where production must be local and low-cost (Bughin et al. 2017).

4.3.4 Transportation and Warehousing

AI has played a significant role in the warehouse by better optimization of assets and processes and designing the best teams, i.e., people and robots. The automation process has taken a big stride because of AI technologies. Robotics one of the advanced branches of AI has taken a central role in the warehouse (Bughin et al. 2017). Advances in technologies in object recognition and semantic segmentation have transformed the behavior of the robots, particularly in the context of how they recognize the properties of the materials and objects they interact with. The new AI-enhanced, camera-equipped robots are trained to recognize empty shelf space. This leads to a dramatic speed advantage over conventional methods of picking objects (Bughin et al. 2017, Martin and Leurent 2017). AI has also been used to correctly identify an object and its position. This enables robots to handle objects without requiring the objects to be in fixed, predefined positions. Ocado, the UK supermarket, uses one of the AI platforms in its retailer's warehouse, where robots steer thousands of product-filled bins over a maze of conveyor belts and deliver them to human packers just in time to fill shopping bags (Dale 2018). Similarly, other robots whisk the bags to delivery vans whose drivers are guided to customers' homes by the best route based on traffic conditions and weather (Bughin et al. 2017). AI-enhanced logistics robots are also able to integrate disturbances in their movement routines via an unsupervised learning engine for dynamics. This capability leads to more precise makeovers and overall improved robustness of processes (Webster et al. 2019). Utility companies use AI for the maintenance of their extensive electrical grids. Using data from sensors, drones, and other hardware, machine learning applications help grid operators avoid decommissioning assets before their useful lives have ended, while simultaneously enabling them to perform more frequent remote inspections and maintenance to keep assets working well (Bughin et al. 2017). Using AI one European power distribution company reduced its cash costs as high as 30% over five years by replacing power transformers. AI is also enabling "preventive maintenance" as well.

Therefore, in a production unit where multiple machines are used, it will indicate possible failure (Bughin et al. 2017, Martin et al. 2017).

4.3.5 Smart Retailing

AI has enabled retailers to increase both the number of customers and the average amount they spend by creating personal and convenient shopping experiences. Retailers now know more about what their shoppers want even before shoppers themselves (Deb et al. 2018). AI forecasts from patterns and volumes of data, i.e., previous transactions, weather forecasts, social media trends, shopping patterns, online viewing history, facial expression analysis, seasonal shopping patterns, etc. (Fildes et al. 2018, Burgess 2018). The best examples are Amazon, Hulu, Netflix, etc. Similarly, a European retailer improved its earnings before interest and taxes (EBIT) by 1 to 2 per cent by using a machine learning algorithm to anticipate fruit and vegetable sales. The company automatically orders more products based on this forecast to maximize turnover and minimize waste. A German company, i.e., Otto cut surplus stock by 20 per cent and reduced product returns by more than two million items a year, using deep learning to analyse billions of transactions and predict what customers will buy before they place an order (Burgess 2018). AI technologies help retailers predict future store performance when expanding their physical footprints. Now retailers optimize their storage space and location using AI. Another important aspect of the retail industry is merchandising. AI helps in merchandising, with opportunities to improve assortment efficiency. Using geospatial and statistical modelling, they predict and minimize their stock. Amazon has embedded AI at the core of its operations. In the retailer's warehouse in Seattle, machine learning algorithms steer thousands of products over a maze of conveyor belts and deliver them to humans just in time to fill shopping bags. Other robots whisk bags to delivery vans whose drivers are guided by an AI application that picks the best route based on weather and traffic conditions (Fildes et al. 2018, Burgess 2018).

4.3.6 Customer Satisfaction

Recently more focus has been given to "user experience", i.e., creating a richer, more tailored, and more convenient experience for the user. Today's business is all about making every customer feel special and welcome, which is not an easy task. This used to be difficult and expensive and was often reserved for only the most lucrative clients. AI technologies like computer vision and machine learning have changed it completely. For example, if a regular supermarket shopper puts a bunch of bananas in his cart, cameras or sensors could relay the information to an AI application that would have a good idea of what the shopper likes based on previous purchases. The app could then, via a video screen in the cart, suggest that bananas would be delicious with a chocolate fondue, which the purchase history suggests the shopper likes, and remind the shopper of where to find the right ingredients (Mortimer and Milford 2018, Bughin et al. 2017). Amazon has built a retail outlet in Seattle that allows shoppers to take food off the shelves and walk directly out of the store without stopping at a checkout kiosk to pay (Metz 2018, Bughin et al. 2017). The store called Amazon Go relies on computer vision to track shoppers after they swipe into the store and associate them with products taken from shelves. When shoppers leave, Amazon debits their accounts for the cost of the items in their bag and emails them a receipt (Metz 2018, Bughin et al. 2017). Delivery through drones is now a reality. Since Amazon successfully delivered a pilot delivery in rural England in 2016 there a surge in this area. Google partnered with Chipotle to deliver burritos at Virginia Tech, and Domino's Pizza with Flirtey completed a commercial delivery of pizzas in New Zealand (Druehl et al. 2018). UPS has partnered with drone company Zipline and governmental organizations in Africa to coordinate emergency medical supplies delivery (such as blood) in Rwanda (Druehl et al. 2018). Amazon now routinely gathers data from drones during home delivery to target future purchases. From healthcare to education to transportation in every sector, AI is providing the ideal tools for operation management (Druehl et al. 2018).

As there is a very limited number of studies on the role of AI in the supply chain and manufacturing sector, especially in the jewellery sector. So, the basic objective behind this research is to analyse the impact of technology on the jewellery manufacturing sector and the overall supply chain.

4.4 Research Hypothesis

H1: There is a significant positive relationship between AI and demand forecast, supplier relations, Production, Transportation and warehousing, Smart retailing, Customer Satisfaction

4.5 Research Methodology

This research includes both primary and secondary data. Primary data is collected through questionnaires and secondary data is collected from research papers, books, trade journals, etc. Non-probabilistic sampling was the most appropriate for this research. The study has been conducted using many manufacturing companies from all over the country. These companies include some major names from the field of Jewellery manufacturing and wholesaling. The survey was conducted by sending out the questionnaire designed specifically for this research to the selected companies who are in the field of manufacturing and maintain their supply chain effectively. The questions were designed in such a way that the respondents could answer on a Likert scale from 1 to 5. Respondents were asked to give their ratings to the questions accordingly. The research conducted includes a total of 53 companies as respondents. Although this is a non-probabilistic sample and the number of respondents was low, its respondents are legitimate and can provide accurate data for conducting the research. To perform the statistical analysis, the data that has been collected using the questionnaire designed, the IBM SPSS Software has been used. The questionnaire includes six factors, that influence the company's supply chain and growth over time. Correlation And Regression tests have been conducted in this study. The tests have been conducted upon pairs of relevant questions.

4.6 Data Analysis

According to all the correlation tables that have been considered using the SPSS software, it can be analysed that all the pairs of data have a moderate dependency on each other and the significance level is less than 0.05 shows that there was enough evidence in the case of this population to be the result of the Pearson Correlation to be true. The Adjusted R square values in the above tables show the percentage of the dependant variables that have been explained by the interdependent variable, i.e., AI. In the above ANOVA tables, the data shows the significant value in all of them to be less than 0.05 hence the data is significant enough to be considered for the outcome. This gives us the conclusion that there is a relationship between all the factors that we have considered. AI is mainly responsible for all these dependent variables like demand forecast, supplier relations, production, transportation and warehousing, smart retailing, Customer satisfaction, etc.

4.7 Conclusion

AI has rooted itself very well in today's businesses. Not only is it limited to the manufacturing and supply chain processes, but it has also affected every process of various businesses and corporates. AI empowers various important factors of a business and it might have already rooted itself way inside our businesses but still, there is a lot of scope in many fields where AI can be used to change the face of our businesses helping us a lot. In the case of Jewellery specifically, the Indian jewellery industry has been showing great improvements with the help of AI. There have been improvements when it comes to designing jewellery and its manufacturing due to AI-powered machinery and processes because more precision has been set in these processes. Incomes have been improved and ultimately and most

Table 1: Descriptive Statistics.

Descriptive Statistics			
	Mean	Std. Deviation	N
Demand forecast	3.35	.955	54
Customer satisfaction	3.25	1.036	53
Transportation & Warehousing	3.25	1.036	53
Supplier Relations	3.56	1.040	54
Production	3.39	.878	54
Smart Retailing	3.58	.908	53
AI	3.37	1.015	54

Table 2: Model Summary.

Model Summary				
Model	R	R Square	Adjusted R Square	Std. Error of the Estimate
1	.506[a]	.256	.241	.832
a. Predictors: (Constant), AI				

Table 3: ANOVA.

ANOVA[a]						
Model		Sum of Squares	df	Mean Square	F	Sig.
1	Regression	12.347	1	12.347	17.851	< .001[b]
	Residual	35.967	52	.692		
	Total	48.315	53			
a. Dependent Variable: Demand Forecast						
b. Predictors: (Constant), AI						

Table 4: Coefficients.

Coefficients[a]								
		Unstandardized Coefficients		Standardized Coefficients		Sig.	95.0% Confidence Interval for B	
Model	B	Std. Error	Beta		t	Lower Bound	Upper Bound	
1	(Constant)	1.749	.396		4.418	< .001	.955	2.543
	AI	.476	.113	.506	4.225	< .001	.250	.701
a. Dependent Variable: Demand Forecast								

importantly customer satisfaction has improved due to this incorporation of AI in various processes of a business.

4.8 Practical Implications and Future Scope

Supply chain managers can utilise data to embed the required degree of resilience in their supply chains by considering the proposed framework elements and phases. There is a need for more studies to examine SC performance impacts of AI particularly in the relationship between autonomous AI manufacturing and operating performance. There is a need for additional research focusing on different AI applications such as autonomous AI-powered manufacturing, load forecasting, and vehicle scheduling. Future research could examine how the resources required change across periods as operations become more autonomous, i.e., pre- and post-implementation. Finally, with SCs continually evolving and marketplaces competitive and dynamic, there is a need to bridge the advances within operational research techniques, such as mathematical optimisation and modelling, simulation theory and control, stochastic programming, neural networks with operations management techniques to better understand how synergies and symbiotic SCs can evolve. Future research could investigate how these two disparate research paradigms could intertwine to reduce risk, increase collaboration and performance, and result in SCs powered by technology yet driven by people.

References

Abbasi, B., T. Babaei, Z. Hosseinifard, K. Smith-Miles, and M. Dehghani. 2020. Predicting solutions of large-scale optimization problems via machine learning: A case study in blood supply chain management. *Computers & Operations Research* 119: 104941.

Adam, F., and B. Fitzgerald. 2000. The status of the information systems field: historical perspective and practical orientation'. *Information Research* 5: 4.

Ahmad, S., and R. G. Schroeder. 2003. The impact of human resource management practices on operational performance: recognizing country and industry differences. *Journal of Operations Management* 21(1): 19–43.

Akter, S., S. F. Wamba, A. Gunasekaran, R. Dubey, and S. J. Childe. 2016. How to improve firm performance using big data analytics capability and business strategy alignment?" *International Journal of Production Economics* 182(C): 113–131.

Ali, M. M., M. Z. Babai, J. E. Boylan, and A. A. Syntetos. 2017. Supply chain forecasting when information is not shared. *European Journal of Operational Research* 260(3): 984–994.

Amabile, T. M. 1983. The social psychology of creativity: a componential conceptualization. *Journal of Personality and Social Psychology* 45(2): 357–376.

Anderson, J. C., and D. W. Gerbing. 1988. Structural equation modeling in practice: a review and recommended two-step approach. *Psychological Bulletin* 103(3): 411–423.

Anshari, M., M. N. Almunawar, S. A. Lim, and A. Al-Mudimigh. 2019. Customer relationship management and big data enabled: personalization & customization of services. *Applied Computing and Informatics* 15(2): 94–101.

Bag, S., L. C. Wood, L. Xu, P. Dhamija, and Y. Kayikci. 2020. Big data analytics as an operational excellence approach to enhance sustainable supply chain performance. *Resources, Conservation and Recycling* 153: 1–11.

Baihaqia, I., and A. S. Sohalb. 2013. The impact of information sharing in supply chains on organisational performance: an empirical study. *Production Planning & Control* 24(8-9): 743–758.

Banville, C., and M. Landry. 1989. Can the field of MIS be disciplined?" *Communications of the ACM* 32(1): 48–60.

Barley, S. R. 1983. Semiotics and the study of occupational and organizational cultures. *Administrative Science Quarterly* 28(3): 393–413.

Barney, J. 1991. The firm resources and sustained competitive advantage. *Journal of Management* 17(1): 99–121.

Barton, D., and D. Court. 2012. Making advanced analytics work for you. *Harvard Business Review* 90(10):78–83.

Baryannis, G., S. Dani, and G. Antoniou. 2019a. Predicting supply chain risks using machine learning: the trade-off between performance and interpretability. *Future Generation Computer Systems* 101: 993–1004.

Baryannis, G., S. Validi, S. Dani, and G. Antoniou. 2019b. Supply chain risk management and artificial intelligence: state of the art and future research directions. *International Journal of Production Research* 57(7): 2179–2202.

Baryannis, G., I. Tachmazidis, S. Batsakis, G. Antoniou, M. Alviano, and E. Papadakis. 2020. A generalised approach for encoding and reasoning with qualitative theories in answer set programming. *Theory and Practice of Logic Programming* 20(5): 687–702.

Baur, C., and D. Wee. 2015. Manufacturing's Next Act. Accessed February 8, 2021. https://www.mckinsey.com/business-functions/operations/our-insights/manufacturings-next-act.

Benbarrad, T., M. Salhaoui, S. B. Kenitar, and M. Arioua. 2021. Intelligent machine vision model for defective product inspection based on machine learning. *Journal of Sensor and Actuator Networks* 10(1): 7.

Ben-Daya, M., E. Hassini, and Z. Bahroun. 2019. Internet of things and supply chain management: a literature review. *International Journal of Production Research* 57(15-16): 4719–4742. doi:10.1080/00207543.2017.1402140.

Bentler, P. M. 1990. Comparative fit indexes in structural models. *Psychological Bulletin* 107(2): 238–246.

Berns, M., A. Townend, Z. Khayat, B. Balagopal, M. Reeves, M. S. Hopkins, and N. Kruschwitz. 2009. The business of sustainability: what it means to managers now. *MIT Sloan Management Review* 51(1): 20–26.

Beugelsdijk, S., N. G. Noorderhaven, and C. Koen. 2009. A dyadic approach to the impact of differences in organizational culture on perceived relationship performance. *Industrial Marketing Management* 38(3): 312–323.

Bienhaus, F., and A. Haddud. 2018. Procurement 4.0: factors influencing the digitisation of procurement and supply chains. *Business Process Management Journal* 24(4): 965–984.

Brintrup, A., J. Pak, D. Ratiney, T. Pearce, P. Wichmann, P. Woodall, and D. McFarlane. 2020. Supply chain data analytics for predicting supplier disruptions: a case study in complex asset manufacturing. *International Journal of Production Research* 58(11): 3330–3341.

Bordeleau, F. E., E. Mosconi, and L. A. De Santa-Eulalia. 2020. Business intelligence and analytics value creation in industry 4.0: A multiple case study in manufacturing medium enterprises. *Production Planning & Control* 31(2-3): 173–185.

Bortolotti, T., S. Boscari, and P. Danese. 2015. Successful lean implementation: organizational culture and soft lean practices. *International Journal of Production and Economics* 160: 182–201.

Brandon-Jones, M., and K. Kauppi. 2018. Examining the antecedents of the technology acceptance model within e-procurement. *International Journal of Operations & Production Management* 38 (1): 22–42.

Bughin, J., E. Hzan, S. Ramaswamy, M. Chui, T. Allas, P. Dahlstrom, N. Henke, and M. Trench. 2017. Artificial intelligence: The next digital frontier? McKinsey Global Institute.

Burgess, A. 2018. AI in Action. In: *The Executive Guide to Artificial Intelligence.* Palgrave Macmillan, Cham.

Cadden, T., G. Cao, Y. Yang, A. McKittrick, R. McIvor, and G. Onofrei. 2020. The effect of buyers' socialization efforts on the culture of their key strategic supplier and its impact on supplier operational performance. *Production Planning & Control.* doi:10.1080/0953 7287.2020.17855.

Correa, J. S., M. Sampaio, R.C. Barros, and W.C. Hilsdorf. 2020. IoT and BDA in the Brazilian future logistics 4.0 scenario. *Production* 30: e20190102.

Dale M. 2018. Automating grocery shopping. *Imaging and Machine Vision Europe* 85: 16.

Dash, R., M. McMurtrey, C. Rebman, and U. K. Kar. 2019. Application of artificial intelligence in automation of supply chain management. *Journal of Strategic Innovation and Sustainability* 14(3).

Davenport, T. H. 2018. The AI Advantage: How to Put the Artificial Intelligence Revolution to Work.

Deb, S. K., R. Jain, and V. Deb. 2018. Artificial Intelligence—Creating Automated Insights for CustomerRelationship Management. 8th International Conference on Cloud Computing, Data Science &Engineering (Confluence).

Dominguez, C., B. Ageron, G. Neubert, and I. Zaoui. 2010. Inter-organizational strategic alignments in a jewelry supply chain using RFID: a case study Supply chain performance: collaboration, alignment and coordination. - London: ISTE, ISBN 1-84821-219-4. - 2010, p. 141–169.

Druehl, C., J. Carrillo, and J. Hsuan. 2018. Innovation and Supply Chain Management Technological Innovations: Impacts on Supply Chains. Springer.

Fildes, R., P. Goodwin, and D. Önkal. 2018. Use and misuse of information in supply chain forecasting of promotion effects. *International Journal of Forecasting.*

Kannabiran, G. and S. Bhaumik. 2005. Corporate turnaround through effective supply chain management: The case of a leading jewellery manufacturer in India. *Supply Chain Management* 10(5): 340–348.

Klumpp, M. 2018. Automation and artificial intelligence in business logistics systems: human reactions and collaboration requirements. *International Journal of Logistics Research and Applications* 21(3).

Kusiak, A. 2018. Smart manufacturing, Volume 56. *International Journal of Production Research,* 508–517.

Martin, C., and H. Leurent. 2017. Technology and Innovation for the Future of Production: Accelerating Value Creation. WEF.

Mittal, S., M.A. Khan, J.A. Purohit, K. Menon, D. Romero, and T. Wuest. 2020 A smart manufacturing adoption framework for SMEs. *International Journal of Production Research* 58: 5, 1555–1573.

Metz, R. 2018. Amazon's cashier-less Seattle grocery store is opening to the public. MIT tech review.

Momme. 2002. Framework for outsourcing manufacturing: Strategic and operational implications. *Computers in Industry* 49(1): 59–75 ·

Mortimer, G., and M. Milford. 2018, August 31. When AI meets your shopping experience it knows what you buy – and what you ought to buy. The Conversation.

Raman, S., N. Patwa, I. Niranjan, U. Ranjan, K. Moorthy, and A. Mehta. 2018 Impact of big data on supply chain management. *International Journal of Logistics Research and Applications* 21: 6, 579–596.

Webster, C., and S. H. Ivanov. 2019, February 17. Robotics, artificial intelligence, and the evolving nature of work. In: George, B., and Paul, J. (eds.). *Business Transformation in Data DrivenSocieties, Palgrave-MacMillan* (Forthcoming). Retrieved from SSRN:https://ssrn.com/abstract=3336104.

CHAPTER 5

Impact of Blockchain in Creating a Sustainable Supply Chain

Piyusha Nayyar and *Pratibha Garg**

Contents

5.1 Introduction

The awareness and concern about the impact of human activities on the environment, across the globe are on the rise (NASA 2021). The Paris Climate Accord and Kyoto protocol offered a viable framework of a 'carbon credit system' to mitigate greenhouse gas emissions using an incentivized market mechanism. Meeting sustainability compliance and consumer awareness about products with low carbon footprints, are driving organisations to minimize carbon emissions. Organisations are trying to minimize carbon emissions across their supply chains using a carbon credit system. However, the lack of standardisation in the

Amity School of Business, Amity University, Uttar Pradesh, Noida.
* Corresponding author: pgarg1@amity.edu

measurement of carbon emission, high cost of emission measurement and non-uniform pricing of carbon credits in different geographies and maintaining transparency, are the few challenges.

Blockchain is one of the disruptive technologies that comes with unique features like immutable data, smart contracts, and peer-to-peer networks (Treiblmaier 2018). Therefore, blockchain can be a befitting technology for building sustainable supply chains with minimizing overall carbon footprint (Kshetri and Loukoianova 2019) blockchain technology can also help in the authentication of sustainability audits and certification.

A sustainable supply chain is a strategic initiative that achieves operational excellence, reduces waste and generates low carbon emissions (Paliwal et al. 2020). The supply chain is comprised of upstream and downstream network participants. According to Shaw et al. (2012), the operation is responsible for just 19 percent of carbon emissions in a supply chain network. The balance of 81 percent carbon emission is generated by upstream and downstream network participants. Therefore, the prime objective of the sustainable supply chain is to reduce overall carbon emissions across the supply chain network (Theiben and Spinler 2014). The sustainable supply chain can be achieved through having a low carbon footprint at every stage, i.e., by low carbon emission product design, low emission at supplier's processes, production process and low emission logistic and distribution (Fuchs 2019).

Many studies have reviewed the impact of blockchain technology on sustainability and supply chains but still knowledge and understanding of its potential remains limited.

This chapter starts with a discussion on the origin and concept of blockchain technology, challenges in building sustainable supply chains and how these challenges can be handled using blockchain technology. The study also talks about how the challenges in audits and certification of sustainable practices can be handled using blockchain and the conclusion.

5.2 Review of Literature

5.2.1 Blockchain

Blockchain technology was introduced in 2008. In the beginning, blockchain technology was considered to be a synonym for the cryptocurrency Bitcoin (Nakamoto and Bitcoin 2008). But, over time, it is realized that the blockchain has unique features and qualities which has the potential to make it a disruptive technology that can be used in various sectors of industries (Pilkington 2016). Blockchain has attracted various industries where it can be useful like healthcare, food industry, consumer goods, Manufacturing, Government Services, and many more in addition to the financial sector. Another interesting application of Blockchain Technology is in building Sustainable Supply Chains where it can act as a game changer.

Blockchain is referred to as a decentralised, constantly increasing list of records known as blocks, throughout a peer-to-peer network which is connected and protected by cryptography (Attaran and Attaran 2007). Blockchain is also called distributed ledger technology which sustains a chronological set of transactions. It's a network that can transfer and accumulate data. DTL retains a chain of data called 'blocks' consisting of unaltered and historic data (Bashir 2017). Each time a new block is generated and included in the chain. Thereafter the unique IDs, transactions, members' addresses etc of the prior block are recorded in the new block. Each successive block holds a unique ID of the group origin of the prior block in the chain certifying that the data is historic and unaltered.

Broadly blockchain is classified into three categories—Public blockchain, Private blockchain and Hybrid blockchain (Wang et al. 2017). The distributed ledger technology holds some unique features that can transform the entire business ecosystem such as immutability, transparency, quicker transaction, consensus and decentralization. As per the Gartner analysis report for the year 2020, blockchain technology is amongst the top ten emerging technology trends (Gartner 2020). Most of the solutions originated by blockchain are in the experimental stage or with partial production.

5.2.2 Sustainable Supply Chains

In our sustainability commitments, several initiatives have been taken by the businesses, regulators and even countries to adopt sustainable practices like reduction of carbon emission in their business processes, for example, carbon emission target, the organisations are required to measure carbon emission throughout the entire supply chain cycle (Theiben and Spinler 2014).

The Paris Climate Accord which was accepted by more than 196 countries, established a framework for measuring and controlling carbon emissions by setting standards for each country or determined nationally linked to their growth objectives (UNFCCC 2021). The individual countries at the national level had options to (a) Specify a limit for carbon emission for the companies (b) impose a carbon tax which the companies have to pay according to the amount of carbon emission generated by them (c) introduce a carbon emission trading scheme. Option (c), i.e., carbon emission trading scheme has become more acceptable as it incentivises the low carbon producing companies and allows time for the others to reduce their carbon footprint gradually.

Kyoto protocol established the 'Carbon Credit System'. The carbon credit system framework awards carbon credits to organisations or entities that exceed their carbon targets by reducing carbon emissions (Chan 2009). One carbon credit is given for the reduction of ten tons of carbon dioxide or equivalent greenhouse gas emissions. Alternatively, carbon credits can be produced by implementing projects which involve activities that reduce carbon emissions. Some of the activities that reduce emissions include planting trees, installing renewable energy generation equipment/infrastructure or encouraging the usage of public transport, etc. (Correia et al. 2013). These carbon credits can be traded in the market. Organisations that cannot limit their carbon emission to their quota assigned by the government/regulator, can buy these carbon credits from the other organisations that earned carbon credits due to their lower emission or through emission reduction activities. In the long run, these organisations can improve technology, product design or processes to reduce the emission. According to a World Bank report, the present market size of carbon credits is close to US$ 50 billion and is expected to reach over 185 billion US$ over the next decade (World Bank Report 2022).

Using the carbon credit system, organisations can overall reduce carbon emissions for their products by building synergies across the supply chain (Shaw and Shankar 2012). For example, a Laptop manufacturer like Dellor HP, requires components like Central Processor Unit, Memory, LCD Screen, Control PCB and Housing Cabinet. Generally, these components are produced by independent organisations as a part of the supply chain for Laptop manufacturers. Some of these supply chain participants might have lower carbon emissions than their assigned quota and others might have higher emissions than their assigned quota. By leveraging the carbon credit system, such laptop manufacturers can reduce overall carbon emissions per Laptop.

5.3 Research Methodology

A secondary research methodology has been used for this research. An extensive review of literature has been carried out on blockchain technology, low carbon emission supply chains and the integration of blockchains in the supply chain. There are many recent studies available on the blockchain and its applications but most of the studies focus on the engineering and technological perspectives of the blockchain. Only a few studies have attempted to bring out the applications of blockchain from the perspective of developing low carbon footprint supply chains, use of blockchain technology audit and certification which are commonly used for sustainable practices and even in the administration of carbon credit market.

5.4 Challenges in Supply Chain

Challenges in Low Carbon Emission Supply Chains

Cost: The cost of measuring and tracking the emission is high. This cost is mostly due to efforts in coordinating with various parties to obtain information which is paid for or not easily accessible.

Standardisation: There is no standardisation in measuring carbon emissions. There is no standard method or technology for measuring carbon emissions globally. Each country/organisation has deployed non-standardised methods to measure the footprint which makes it difficult to compare the carbon footprint of similar organisations in a different geography.

Differences across different countries: Differences in measuring and pricing and terms of carbon credits across different countries.

Challenges in Audits and Certification

Prone to Fraud: These audits and certification reports are generally issued as PDF documents which are prone to a temper which can be the removal of any non-compliance or violation or altering the score for some parameters. For example, if a retailer wants to verify the correctness of the report or certificate, it has very limited means to verify the report. Therefore the authenticity of the certification and reports is a challenge.

Additional overhead: There are a large number of transaction audit certificates in the supply chain (One per transaction). Managing such a large number of certificates adds another overhead cost for the supply chain participants.

Delay in receiving certificates and reports: There might be a delay in receiving audit certificates by the retailers. For example, a retailer buys a textile made from premium cotton. But he may not get the certificate along with the shipment.

Challenges in the Carbon Credit Market

Two types of carbon credit markets are operational (1) voluntary markets where carbon offsets are exchanged between buyer and seller directly. The voluntarily created offsets are called VER (voluntary emission reduction). (2) The compliance market in which the organisations need to reduce the emission to capped levels as per government regulations or they have to purchase the offsets. These offsets are called CER (Certified Emission Reduction). The major difference between VER and CER is that CER is regulated by a third-party certifying body whereas VER is not. Both

Voluntary and compliance markets have challenges due to fragmented implementations. These also lack cross-market valuations and non-uniform carbon credit valuation of projects. Another big challenge in the present carbon credit market is the possibility of double counting or multiple parties claiming the credit of the same offset. A carbon offset registry is used to track the offset with a serial number and the owner's details. If the offset is sold to another stakeholder, the ownership is transferred accordingly in the registry. If the offset is used by the owner to settle its emission, the offset retires and that serial number offset cannot be resold. However, in the voluntary market, many agencies develop registries including government, private organisations, and Non-profit organisations. Therefore it has the potential for double entry. Further, the database used for maintaining the registry is not temper proof. There are possibilities of manipulation in the registry database. Even the cost of transaction is quite high as brokers and agents are involved in the trade which adds 5 to 8% additional cost.

5.5 Used Cases on Blockchain for Sustainable Practices

Blockchain in 'low carbon emission' for sustainable practices: Blockchain technology is the perfect candidate for providing a solution to the above challenges. As Blockchain technology comes with inherent qualities of immutable data, quick authentication and smart contracts, it can offer standardised, accurate and easily accessible information for carbon emission measurement across supply chains (Saraji and Borowczak 2021). The Laptop manufacturer can set up a blockchain with other supply chain participants as permission members. The supply chain members can define an agreed standard for emission measurement, terms and pricing for carbon credit systems and use smart contracts to implement the same (Christidis and Devetsikiotis 2016). This will address the challenges of standardisation in measurement and pricing and terms of the carbon credit system. As the blockchain system can be integrated with SCADA and other ERP solutions, which can simplify and reduce the cost of measurement. Therefore blockchain can help in creating a sustainable supply chain platform where all supply chain participants like suppliers, manufacturers, distributors & retailers can operate with transparency, accountability and efficiency (Vranken 2017). Figure 1 shows that blockchain enables the supply chain for laptop manufacturing.

Blockchain in 'Audits & Certifications' for sustainable practices: In almost every supply chain, auditors are assigned to audit the entities of supply chain participants and provide audit reports and certifications to participants who need them. These audits are of broadly two types, i.e., Site Audits & Transaction Audits. Site Audits are conducted at the facility

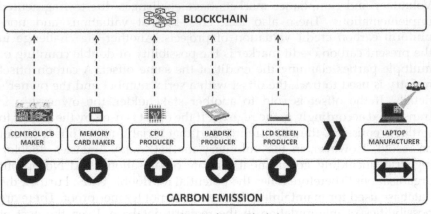

Figure 1: Site Audits in Blockchain.

of the entity against a set of standards which are industry specific. Site audits generally include

- Sustainable practices implemented for water & land conservation.
- Sustainable practices are adopted in the production process to conserve the environment.
- Working conditions for the humane.
- To ensure from abuse of child labour.

Transaction Audits are the ones which are used to certify the exchange of goods between two parties. This audit is carried out by a third-party auditor which certifies the exchange of goods of certain qualities with attributes. For example, a farmer sold organic cotton of premium fibre quality to a spinning mill. The auditor certifies the transaction of exchange of premium quality cotton between the farmer and the spinning mill.

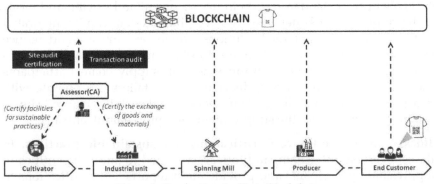

Figure 2: Transaction Audits in Blockchain.

Here blockchain technology can be really handy to overcome these challenges. These site and transaction audit reports and certificates can be recorded in a blockchain network which is accessible to all the supply chain participants instantly. Even a QR code on the end product (e.g., a Shirt in this example) can instantaneously provide relevant information to the consumer. The above diagram depicts a blockchain-enabled supply chain for cotton textiles.

Blockchain in the administration of Carbon Credit Market

The challenges related to the carbon credit market can be addressed using blockchain technology which is temper proof, and peer-to-peer where no broker or agent is required (Banerjee 2018). Using a tokenised offset based on blockchain technology can be used to eliminate the possibilities of a double count or multiple claims for the same offset.

5.6 Conclusion

Blockchain technology will help in building a sustainable supply chain. It can address the challenges of standardisation in the measurement of carbon emission across supply chains which would help in reducing the overall carbon footprint across the supply chains. Blockchain can be extremely useful in the carbon credit market by eliminating double counting and multiple claims for the same offset. There are tremendous opportunities for use of blockchain technology in addressing the challenges of authentication and minimizing the frauds in sustainability audits and certification which will benefit all supply chain participants and consumers.

References

Attaran, M., and S. Attaran. 2007. Collaborative supply chain management. *Business Process Management Journal*.

Banerjee, A. 2018. Re-engineering the carbon supply chain with Blockchain technology, Whitepaper Infosys.

Bashir, I. 2017. Mastering blockchain. Packt Publishing Ltd.

Chan, M.-K. 2009. Carbon Management. A Practical Guide for Suppliers; University of Cambridge: Cambridge, UK.

Christidis, K., and M. Devetsikiotis. 2016. Blockchains and smart contracts for the internet of things. *Ieee Access*. 2016 May 10; 4: 2292–303.

Correia, F., M. Howard, B. Hawkins, A. Pye, and R. Lamming. 2013. Low carbon procurement: An emerging agenda. *J. Purch. Supply Manag*. 19: 58–64.

Fuchs, S. J., D. N. Espinoza, C. L. Lopano, A.-T. Akono, and C. J. Werth. 2019. Geochemical and geomechanical alteration of siliciclastic reservoir rock by supercritical CO_2-saturated brine formed during geological carbon sequestration. *Int. J. Greenh. Gas Control* 88: 251–260.

Gartner. 2020. Gartner's report: Hype cycle for Blockchain Technologies.

Kshetri, N., and E. Loukoianova. 2019. Blockchain Adoption in Supply Chain Networks in Asia. *IT Prof.* 21: 11–15.

Nakamoto, S., and A. Bitcoin. 2008. A peer-to-peer electronic cash system. Available via Bitcoin. https://bitcoin. org/bitcoin. Accessed 15 Feb 2021.

NASA. (n.d.) The Intergovernmental Panel on Climate Change. Available online: https://climate.nasa.gov/causes/ (accessed on 22 November 2021).

Paliwal, V., S. Chandra, and S. Sharma. 2020. Blockchain technology for sustainable supply chain management: a systematic literature review and a classification framework. *Sustainability* 2020 Jan; 12(18): 7638.

Pilkington., M. (eds.). 2016. Blockchain technology: principles and applications. *Research Handbook on Digital Transformations*. SSRN 2662660.

Saraji, S., and M. Borowczak. 2021. A Blockchain-based Carbon Credit Ecosystem, White paper, University of Wyoming.

Shaw, K., R. Shankar, S. S. Yadav, and L. S. Thakur. 2012. Supplier selection using fuzzy AHP and fuzzy multi-objective linear programming for developing low carbon supply chain. *Expert Syst. Appl.* 39: 8182–8192.

Theißen, S. and S. Spinler. 2014. Strategic analysis of manufacturer-supplier partnerships: An ANP model for collaborative CO2 reduction management. *Eur. J. Oper. Res.* 233: 383–397.

Treiblmaier, H. 2018. The impact of the blockchain on the supply chain: A theory-based research framework and a call for action. *Supply Chain Manag. Int. J.* 23: 545–559.

UNFCCC. The Paris Agreement. 2021. Available online: https://unfccc.int/process-and-meetings/the-paris-agreement/the-paris-agreement (accessed on 23 November 2021).

Vranken, H. 2017. Sustainability of Bitcoin and Blockchains, Elsevier Current Opinion in Environmental Sustainability, 28: 1–9, October, 2017.

World Bank. State and trends of carbon pricing 2021. Available online: https://openknowledge.worldbank.org/handle/10986/35620 (accessed on 21 January 2022).

CHAPTER 6

Exploring Adoption of Blockchain Technology for Sustainable Supply Chain Management

Salaj,[1] *Subhrata Das*[2] *and Mohini Agarwal*[3],*

Contents

6.1 Introduction

The Supply Chain is a network of related and mutually dependent organizations working together to control, manage and improve the flow of materials and data from suppliers to end customers (Mentzer et al. 2001).

[1] School of Basic and Applied Sciences, K R Mangalam University, Gurugram-122103.
[2] Department of Operational Research, University of Delhi, Delhi-07.
[3] Independent Researcher, Arizona, United States, 85308.
* Corresponding author: mohini15oct@gmail.com

The period of globalization has forced organizations to give forth the best quality products which are sustainable to the customers wherein supply chain management plays a vital role (Mehta 2004). All organizations are extremally reliant upon the effective supply chain process (Maloni et al. 1997, Fawcett et al. 2008) which can assist firms in having the following significant benefits such as to build better customer relationships and services (Alshurideh et al. 2019), develop a better mechanism for speedy delivery (Ellram et al. 2004), enhanced productivity and business capacities (Sangari et al. 2015), minimises storage and transportation costs, minimizes overall expenses in supply chain (Ghaffariyan et al. 2013), helps in accomplishing delivery of right items to the perfect locations at right time (Moons et al. 2019), upgrades stock administration (Maloni et al. 1997, Fawcett et al. 2008), supporting the effective execution of just-in-time stock models (Wang and Bhaba 2004), helps organizations in adapting to the diversified customer demand spread globally (Maloni et al. 1997, Fawcett et al. 2008), help organizations in limiting waste, driving out costs, and accomplishing efficiencies all through the supply chain process (Marchi et al. 2013).

At present, the manufacturing sector's biggest requirement is to maintain a balance between environmental, social, and business outcomes and to adopt sustainable practices which enable to meet the growing needs with minimum impact on the environment (Braccini et al. 2018, Saberi et al. 2019, Ghobakhloo 2020, Vrchota et al. 2020). Considering the significance of sustainable practices, manufacturers have adopted different tools and approaches, for example, distributed computing, big data analytics, blockchain, artificial intelligence, lean manufacturing, six sigma, and reverse logistics (Shaharudin et al. 2017). Blockchain is one of the rising and quickly developing technologies that can add to the sustainable practice of manufacturing among computerized technologies. Many large financial services firms are committing millions of dollars in research and development to explore and evaluate potential applications, but the technology is also capturing attention in sustainability circles (Treleaven et al. 2017).

The term Blockchain gained attention for the first time due to Bitcoin, the first cryptocurrency (Nakamoto 2008), and its ability to make a trusted and straightforward ledger of exchange data. Nowadays supply chain managers are very hopeful and enthusiastic to use this technology due to its characteristic of immutability and transparency (Saberi et al. 2019). This is the appropriate time to use this technology as the customers are demanding transparency in the supply chain. Before buying any product, customers are keen to know the source of raw material, manufacturing time and all the set agreements signed, and many more. For example, jewelry buyers need confirmation that purchased diamonds are legitimate and not mined from illegal areas of the world; the sender of goods has been interested in tracking the current location and the condition of cargo; organizations want to track the usage cycle, maintenance activities, and operational conditions of the machinery and related parts. The problem of manufacturing and delivery in a sustainable manner gets intensified when supply chains are multi-layered and globally connected. These concerns can be easily managed with the application of blockchain technology with its key components like a shared ledger, smart contract, privacy, trust, and transparency (Treleaven et al. 2017) which makes it imperative to explore the highlights and challenges in the adoption of blockchain technology within the sustainable supply chain.

Blockchain technology promises to significantly change transaction methods by giving a straightforward and permanent record for investigation (Zheng et al. 2017). At present, blockchain applications are limited to a few sectors, among a few companies wherein fundamentally they have been utilized and developed for the finance sector (Treleaven et al. 2017), however, well-known companies and supply chain managers have paid attention and are applying the technology to provide better consumer support and achieve transparency in the system (Treleaven et al. 2017). The purpose of this chapter is therefore to:

1. Provide an overview of blockchain technology to contribute positively, taking reference from what is currently known about blockchain technology in the extant literature and narrowing it to the field of SCM.

2. Present an outline of the various highlights and challenges associated with the adoption of the technology in the SCM.

3. Suggest streams of SCM which can benefit from the application of blockchain technology.

4. Discuss a set of challenges in the adoption of blockchain technology in supply chain management.

The chapter is organized as follows, Section 6.2 provides an overview where a brief introduction to blockchain technology has been provided. Section 6.3 presents the highlights and challenges in the adoption of Blockchain technology in a general sense. In Section 6.4, the application of blockchain technology in supply chain management by some well-known companies has been discussed. Section 6.5 discusses and classifies various challenges for Blockchain technology adoption in Supply chain management. Following it, the conclusion is in Section 6.6 and lastly the list of references.

6.2 Overview of Blockchain Technology

Blockchain is a powerful and flexible evolving technology which is most popular for its utilization in digital currency. Blockchain is a distributed data set that encourages trust, brings down costs associated with every transaction made by organizations, and can change the way companies work. As of now, the use of technology is in different business-and public-areas, for example, in the financial sector for facilitating and tracking transactions, tracking the cargos for timely delivery, automation of various custom clearances, monitoring the quality of products, tracing the raw materials used in the manufacturing of products. As mentioned by (Yli-Huumo et al. 2016, Banerjee 2018) blockchain technology is built on five main pillars:

- Shared Ledger is an important pillar of blockchain, a distributed system consisting of append functionality. Making it a highly secure part of the database which can easily be shared among all participants to ensure visibility and resilience (Al-Saqaf and Seidler 2017, Irannezhad 2020).
- Smart Contracts are one of the aspects of blockchain with its ability of electronic agreements that self-execute according to mentioned prerequisites (Irannezhad 2020).
- Privacy is ensured as all transactions stored in the distributed ledger are reliable, authentic, and verifiable (Irannezhad 2020).
- Trust is particularly important and is ensured in all transactions as the relevant partners validate them (Irannezhad 2020).
- Transparency in the system is ensured as all the records are immutable and all the participants in the network are aware of all the transactions that impact them (Cai and Zhu 2016, Adams et al. 2017, Irannezhad 2020).

Regardless of these key pillars, some other features make blockchain distinctive and promising which include decreased cost (transaction, administrative, operational, processing), improved resilience and system robustness, security, and efficiency (reduced monitoring, enhanced productivity, data accuracy). Although it is a new technology, it has a rich and fascinating history. In 2008, Satoshi Nakamoto published the first ever article on "Bitcoin: A Peer-to-Peer Electronic Cash System" (Nakamoto 2008). A year later the first ever bitcoin transaction occurred between computer scientist Hal Finney and Satoshi Nakamoto. In the years spanning from 2010 to 2020, many significant events have occurred towards the adoption of blockchain technology and bitcoins as the form of currency for the acceptance of donations and various other payments.

During the same period, Pilkington (2016) described the main principles behind blockchain technology and some of its cutting-edge applications. Crosby et al. (2016) mentioned that blockchain makes it a very attractive technology to solve the current financial as well as non-financial industry problems. Ahram et al. (2017) highlighted and focused on the use of blockchain technology in multiple industries such as healthcare and finance, government, and various manufacturing sectors wherein security, scalability and efficiency are most important. Cocco et al. (2017) looked at the challenges and opportunities of implementing blockchain technology across banking, providing food for thought about the potentialities of this disruptive technology. Che et al. (2018) focused on the application of blockchain to the educational sector and explored how blockchain technology can be used to solve certain education problems. Golosova and Andrejs (2018) focused on the modern industries and analyzed how the industries would be impacted by blockchain integration. Yaga et al. (2019) presented the technical outline of blockchain technology which can help readers understand the terminologies, functionalities, and working of blockchain technology. Wang and Min (2020) highlighted the integration of blockchain and energy and what future it holds for energy.

6.3 Highlights and Challenges in the Adoption of Blockchain Technology

Blockchain innovation is creating a whip in logistics and supply chain management (Pournader et al. 2020). One foremost benefit of blockchains is the extent of safety they can provide, and this additionally means that blockchains can offer secure, speedy, and convenient online transactions (Korpela et al. 2017). It takes a couple of minutes, whereas other transaction methods can take many days to complete. There is no interference from financial institutions or authorities' corporations, which is an added advantage as visualized by users. With the use of Blockchain Technology, users can control information and transactions; can have access to complete, consistent, and up-to-date data without accuracy; the history of any transaction can easily be tracked as all the transactions are digitally stamped. The immutability and transparency of transactions make the system secure as the transactions cannot be modified or deleted. Using blockchain technology makes the system resilient to cyber-attacks due to its peer-to-peer nature and multiple copies of the data can be created and damage and loss of sensitive information can be minimized. Band and various other payment systems are using blockchain technology for smoother, efficient, and transparent transactions ensuring that the system and records remain secure (Sarmah 2018).

The application of blockchain in healthcare can/has led to restoring the lost trust between the customer and the healthcare providers as identifying and authorizing information about any medical record of people becomes easier and fraud-free. Due to the inherent property of storing, verifying, and digitally stamping documents, the legal industries have adopted blockchain technology for verifying data and keeping the documents securely. The number of court cases can be significantly reduced by authenticating and verifying facts and figures by using blockchain technology. Not only can this blockchain technology also be used to avoid rigging election results, voters' registration, and their validation but can also assist in creating a public ledger of casted votes to ensure the legitimacy of elections. Big players in the insurance, education, transportation and retail sectors are finding blockchain technology suitable for building trust, reducing costs, and increasing transparency.

The challenges with the adoption and implantation of blockchain are manifold: technical know-how of blockchain technology, figuring out how blockchain can be molded to create value, and advancing methods that can capture early value to facilitate the journey (Uniliver News 2019). Blockchain and cryptography include using public and private keys, and reportedly, there were issues with private keys. One major challenge is if a consumer loses their private key. Scalability restriction (the number of transactions per node is limited) can sometimes take more time than expected in completing several transactions (Sarmah 2018). It also can be hard to append records after it's recorded, that's another big downside of blockchain. Implementation of blockchain technology can be expensive as it is resource intensive. Blockchain is full of complex concepts and processes making it difficult to understand and comprehend by common people leading to its restrictive implementation.

In the blockchain, all the transaction-related information is available publicly which could end up as liability when the data stored in the distributed ledger contains crucial government data or health records of patients such a type of ledger should restrict access to information to all. If the restriction is applied then there is a loss of transparency making the implantation of blockchain technology a matter of concern. Even though the technology is complex and complicacy to understand but several market players have used and even benefitted from the implementation of blockchain technology. Apart from the digital currency, blockchain technology is a new computing and information flow paradigm with broad implications for future development in supply chain management and logistics (Abeyratne and Monfared 2016, Tian 2016, Maurer 2017). In the following section, the focus has been to understand how the technology has benefitted the field of supply chain management.

6.4 Application of Blockchain Technology in Supply Chain Management

Resilience in the supply chain can be built by having a grip on internal processes and data, operational visibility, the ability to anticipate a way forward, maintain and nurture relationships with various supply chain partners, capability to model scenarios that can assist in making faster and better decisions to handle critical situations. Consequently, blockchain decentralization and disintermediation functions (Scott et al. 2017) can result in disruption and support supply chain innovation and reconfiguration in the virtual age. Several market players have successfully implemented blockchain technology and have gained the trust and generated transparency in their systems. It comes as no surprise that the Economist called Blockchain technology The Trust Machine in 2015 (Sristy 2021).

In the second quarter of the year 2018, M&S claimed to be the only national retailer in the UK that can trace every piece of beef it sells, all the way back to every farm and animal from which it is sourced. This showcases the company's capability in generating trust among customers by tracing every bit of beef used (M&S Press Release 2018). In 2018, Swiss food manufacturer Nestlé SA is engaged with a project to test the utilization of blockchain innovation. The key point is to work on the transparency of the overall food-fixing store network, and consequently reinforce buyer trust in sanitation. As per a report in The Wall Street Journal earlier, the 'Food Trust' project includes nine other food industry monsters, including Unilever and Walmart (SCM trends 2019). Further in 2019, they made use of blockchain technology for acquiring more transparency in the food chain. The organization has begun a pilot-based blockchain distributed ledger technology that enables the end-to-end traceability of food.

Beginning of 2019, the Unilever organization used blockchain technology to minimize manual intrusion and avoid arguments between suppliers. Recently, they have been looking for ways of making a reasonable backwater impression, so by adding blockchain technology to furnish transparency and traceability to its palm oil inventory network and working on the maintainability of palm oil production.

Several other instances of Blockchain technology implementation (Sristy 2021).

- The De Beers Group tracked 100 high-value diamonds along its supply chain from mine to retail, creating unprecedented asset-traceability assurance.
- the UAE's state-owned oil company, Abu Dhabi National Oil Company (ADNOC), launched a Blockchain supply chain system pilot program

in collaboration with IBM to track oil from the well to the customers and automate the transactions that take place along the way.

- Circa 2017, global shipping giant Maersk completed the first test of its Blockchain tech to manage cargo. The company developed the TradeLens supply chain platform with IBM to help track cargo ships and containers.
- FedEx is a part of the Blockchain in Transport Alliance (BiTA) and launched a Blockchain-powered pilot program to clarify the data stored on the Blockchain to best resolve customer disputes by helping them receive information in a more streamlined manner.
- Fish supplying firm John West started including codes on the tuna cans that allowed customers to trace the product back to the fishermen.
- In late Jan 2020, Ford Motor Company announced that it would use Blockchain tech to trace cobalt supplies used in electric car batteries to ensure that they get an authentic product to maintain the quality.

With the rising concern in the food supply chain, it is important to develop system applications that can assist in minimizing food fraud and can provide better food traceability which can easily be done with the implementation of blockchain technology as mentioned above. Blockchain implementation has to go through a long process as some projects have already been implemented and some are pilot projects.

Apart from focusing on the implementation of Blockchain in the food supply chain, it can provide benefits to environmental supply chain sustainability (Saberi et al. 2019):

- Keeping track of the product can help in reducing the rework and recall, further reducing the emission of greenhouse gases and resource consumption.
- The use of blockchain technology can be used to ensure that green products are environmentally friendly. Generally, the authenticity of the processing information of green products is unavailable which makes it difficult to verify. Case of the information, if accessible to consumers, they will be more willing to purchase such green products.
- With blockchain implementation, it becomes easier to trace the product footprint of a particular company. This can help customers with the information about their carbon footprint, they can switch to a product with lower carbon footprints. This information can put pressure on companies to restructure their supply chain to reduce carbon emissions.
- The process of recycling can progress with the adoption of blockchain technology. Without rewards, it is difficult to motivate participation in recycling programs, which can be done by using blockchain

technology. The rewards can be offered in terms of cryptographic tokens in exchange for recyclable materials deposited by individuals or companies. Meanwhile, with the absence of blockchain technology, it is difficult to identify and make a contrast between any such initiative's existence. For example, Social Plastic and RecycleToCoin are blockchain technology-based projects to help recycle plastic into rewards (Futurethinkers 2017). This application for a closed-loop supply chain can be applied in the area of the circular economy.

With the application of blockchain technology, fraud can be avoided due to the reliability and transparency of the blockchain. Transparently tracking the product from its origin to the shelf help in curbing the carbon footprint and identifying and eliminating unsustainable practices in the entire supply chain. Any new technology is accompanied by challenges, but as participants in the supply chain, the focus should be on opportunity rather than on threat accompanied by the adoption of blockchain technology in sustainable supply chain management.

6.5 Understanding Challenges to Blockchain Technology Adoption in the Supply Chain

At present, the applicability and how blockchain technology can change and disrupt supply chain design. Let us also understand some of the concerns related to the adoption of blockchain technology. The challenges have been identified based on the literature associated with supply chain systems and are categorized into four groups namely intra-organizational challenges, inter-organizational challenges, system challenges, and external challenges (see Fig. 1).

Intra-organizational challenges are due to the internal activities of the organizations thus these challenges can vary from organization to organization and on the support received from top management.

- The adoption of blockchain technology requires hardware and software which is costly and imposes a financial burden on organizations (Mougayar 2016).
- Lack of management commitment and support commitment obstructs the integrity of sustainability practices (Govindan and Hasanagic 2018).
- The lack of new organizational policies for using technologies would require the transformation of existing organizational culture and would require a substantial change in roles and responsibilities (Mendling et al. 2017).

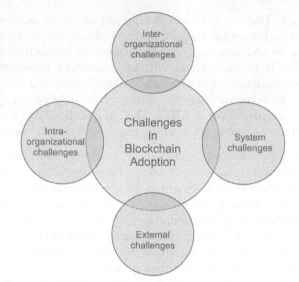

Figure 1: Challenges in Adoption of Blockchain Technology in Supply Chain Management.

- Lack of knowledge and expertise even though the scope of technology is wide there are still limited applications and software developers for the same (Mougayar 2016).
- Difficulty in changing organizational culture as with the adoption of blockchain technology various aspects of the supply chain needs to be modified and this should also be reflected in their vision and mission (Tseng et al. 2015).
- Lack of tools for blockchain technology implementation in the sustainable supply chain as still the technology is only adopted by big market players (Mougayar 2016).

Inter-organizational challenges focus on various partners in the supply chain. A successful supply chain aims to create value for all the partners and stakeholders involved in the supply chain. Sometimes the relationship between various partners becomes perplexing, especially when it comes to adopting newer technology and sustainable practices. With the adoption of blockchain technology, information sharing becomes easier among various partners in the supply chain (Sayogo et al. 2015). Even though sharing information and transparency is the need for attaining a sustainable supply chain, some organizations may not be willing to share a portion of the information which they may find competitive over others. The reluctance to share complete information with all partners within the supply chain limit the benefits of the adoption of blockchain technology in a sustainable supply chain. Following are some of the significant challenges:

- Customers and other participants in the supply chain lack the knowledge and capabilities to adopt sustainable practices and blockchain technology, thus they resist such adoption which can make everything transparent.

- To have a beneficial adoption of technology all the participants in the supply chain should communicate, coordinate, and collaborate well.

- It can not always be a situation in which all partners would be ready to share complete information unless they trust the information would not be misused.

- For a successful supply chain, trust is important and it's important to accommodate the cultural differences of all partners involved in it which is a challenge sometimes.

- Integrating sustainable practices and blockchain technology could be a challenge in supply chain management.

To successfully implement new technology, there comes the need for new IT tools which can be a challenge for all participants in the supply chain. Such system-related challenge forms another class.

- Data security and privacy apprehensions by various participants towards the adoption of blockchain technology (Swan 2015, Mougayar 2016).

- For all the participants to have an advantage from opportunities in the integrated sustainable supply chain they need to have access to complete information, thus having the technology to have access to real-time data can be a challenge.

- Resistance to adopting blockchain technology due to negative perception of the public (Palombini 2017).

- The immutability aspect of blockchain possess a challenge (Tian 2016).

- Immaturity of blockchain technology in terms of several transactions it can handle (Yli-Huumo et al. 2016).

External challenges to the adoption of blockchain technology in sustainable supply chain management originates from external agencies, stakeholders and bodies not directly benefit from the supply chain events. There can be some pressure from different profit or not-for-profit organizations demanding the inclusion of sustainable practices and adopting newer technological practices, which can force organizations to integrate them into their supply chain processes. Following are some of the external challenges faced in supply chain management:

- There are no clearly defined rules and regulations regarding the usage of blockchain technology and their willingness to support sustainable practices (Mangla et al. 2018) becomes a hurdle in integrating sustainable practices and blockchain technology.

- Many organizations face competition and they need to have the assurance that their investment in manufacturing green production, adoption of sustainable practices and acceptance to implement blockchain technology all this is valued by their stakeholders. Implying uncertainty in demand (Kaur et al. 2018) and competition as a challenge.

- To have the benefit of integration of sustainable practices and blockchain technology requires the involvement of various entities and different stakeholders (Mangla et al. 2018).

- Not all organizations are involved in ethical and safe practices which is a hurdle in the growth of the organization.

Discussing and classifying challenges in blockchain technology provides a road towards understanding the new technology and the inclusion of sustainable practices in supply chain networks. These challenges help build theoretical knowledge and can make the organization prepare beforehand.

6.6 Conclusion

In this chapter, a brief discussion about the various highlights and challenges in blockchain technology along with the basics of the technology has been discussed. The various application of blockchain in well-known companies' supply chain management has been reviewed. To broaden the horizon and look towards the adoption of blockchain technology along with sustainable practices, various challenges that may crop up have been theoretically described. There are indeed many opportunities for Sustainable Supply Chain research into Blockchain, including how it will address important Sustainable Supply Chain Transparency challenges such as those related to product counterfeiting, comparisons with other technological implementations, and how its implementation will affect trust and change supply chain relationships, the eventual study of implementation successes (and failures) across supply chains, and so on.

References

Abeyratne, S. A., and R. P. Monfared. 2016. Blockchain ready manufacturing supply chain using a distributed ledger. *International Journal of Research in Engineering and Technology* 5(9): 1–10.

Adams, R., G. Parry, P. Godsiff, and P. Ward. 2017. The future of money and further applications of the blockchain. *Strategic Change* 26(5): 417–422.

Ahram, T., A. Sargolzaei, S. Sargolzaei, J. Daniels, and B. Amaba. 2017. Blockchain technology innovations. In *2017 IEEE Technology & Engineering Management Conference (TEMSCON)*, pp. 137–141. IEEE.

Al-Saqaf, W., and N. Seidler. 2017. Blockchain technology for social impact: opportunities and challenges ahead. *Journal of Cyber Policy* 2(3): 338–354.

Alshurideh, M., N. M. Alsharari, and B. Al Kurdi. 2019. Supply chain integration and customer relationship management in the airline logistics. *Theoretical Economics Letters* 9(02): 392.

Banerjee, A. 2018. Blockchain technology: supply chain insights from ERP. *In Advances in Computers* 111: 69–98. Elsevier.

Braccini, A. M., and E. G. Margherita. 2018. Exploring organizational sustainability of industry 4.0 under the triple bottom line: The case of a manufacturing company. *Sustainability* 11(1): 36.

Cai, Y., and D. Zhu. Fraud detections for online businesses: a perspective from blockchain technology. *Financial Innovation* 2(1): 1–10.

Chen, G., B. Xu, M. Lu, and N.-S. Chen. 2018. Exploring blockchain technology and its potential applications for education. *Smart Learning Environments* 5(1): 1–10.

Cocco, L., A. Pinna, and M. Marchesi. 2017. Banking on blockchain: Costs savings thanks to the blockchain technology. *Future Internet* 9(3): 25.

Crosby, M., P. Pattanayak, S. Verma, and V. Kalyanaraman. 2016. Blockchain technology: Beyond bitcoin. *Applied Innovation* 2(6–16): 71.

Ellram, L. M., W. L. Tate, and C. Billington. 2004. Understanding and managing the services supply chain. *Journal of Supply Chain Management* 40(3): 17–32.

Fawcett, S. E., G. M. Magnan, and M. W. McCarter. 2008. Benefits, barriers, and bridges to effective supply chain management. *Supply Chain Management: An International Journal*.

Futurethinkers. 2017. 7 Ways the Blockchain Can Save The Environment and Stop Climate Change. http://futurethinkers.org/blockchain-environment-climate-change/.

Ghaffariyan, M. R., M. Acuna, and M. Brown. 2013. Analysing the effect of five operational factors on forest residue supply chain costs: A case study in Western Australia. *Biomass and Bioenergy* 59: 486–493.

Ghobakhloo, M. 2020. Industry 4.0, digitization, and opportunities for sustainability. *Journal of Cleaner Production* 252: 119869.

Golosova, J., and A. Romanovs. 2018. The advantages and disadvantages of the blockchain technology. *In 2018 IEEE 6th Workshop on Advances in Information, Electronic aAnd Electrical Engineering (AIEEE)*, pp. 1–6. IEEE.

Govindan, K., and M. Hasanagic. 2018. A systematic review on drivers, barriers, and practices towards circular economy: a supply chain perspective. *International Journal of Production Research* 56(1–2): 278–311.

Irannezhad, E. 2020. Is blockchain a solution for logistics and freight transportation problems? *Transportation Research Procedia* 48: 290–306.

Kaur, H., and S. P. Singh. 2018. Environmentally sustainable stochastic procurement model. *Management of Environmental Quality: An International Journal*.

Korpela, K., J. Hallikas, and T. Dahlberg. 2017. Digital supply chain transformation toward blockchain integration. *In Proceedings of the 50th Hawaii International Conference on System Sciences.*

M&S Press Release. 2018. https://corporate.marksandspencer.com/media/press-releases/2018/m-and-s-raises-the-stakes-with-unrivalled-new-british-beef-traceability-campaign, access date 17th June 2022.

Maloni, M. J., and W. C. Benton. 1997. Supply chain partnerships: opportunities for operations research. *European Journal of Operational Research* 101(3): 419–429.

Mangla, S. K., S. Luthra, N. Mishra, A. Singh, N. P. Rana, M. Dora, and Y. Dwivedi. 2018. Barriers to effective circular supply chain management in a developing country context. *Production Planning & Control* 29, no. (6): 551–569.

Marchi, V. D., E. Di Maria, and S. Micelli. 2013. Environmental strategies, upgrading and competitive advantage in global value chains. *Business Strategy and the Environment* 22(1): 62–72.

Maurer, B. 2017. Blockchains are a diamond's best friend. *Money talks: Explaining How Money Really Works* 215–229.

Mehta, J. 2004. Supply chain management in a global economy. *Total Quality Management & Business Excellence* 15(5-6): 841–848.

Mendling, J., G. Decker, R. Hull, H.A. Reijers, and I. Weber. 2018. How do machine learning, robotic process automation, and blockchains affect the human factor in business process management? *Communications of the Association for Information Systems* 43(1): 19.

Mentzer, J. T., W. DeWitt, J. S. Keebler, S. Min, N. W. Nix, C. D. Smith, and Z. G. Zacharia. 2001. Defining supply chain management. *Journal of Business Logistics* 22(2): 1–25.

Moons, K., G. Waeyenbergh, and L. Pintelon. 2019. Measuring the logistics performance of internal hospital supply chains–a literature study. *Omega* 82: 205–217.

Mougayar, W. 2016. *The Business Blockchain: Promise, Practice, and Application of the Next Internet Technology.* John Wiley & Sons.

Nakamoto, S. 2008. Bitcoin: a peer-to-peer electronic cash system, online, available at: www.bitcoin.org/bitcoin.pdf.

Palombini, M. 2017. The other side of blockchain: we choose what we want to see. *IEEE SA Beyond Standards. Last Modified* 23.

Pilkington, M. 2016. Blockchain technology: principles and applications. *In Research Handbook on Digital Transformations.* Edward Elgar Publishing.

Pournader, M., Y. Shi, S. Seuring, and S. C. L. Koh. 2020. Blockchain applications in supply chains, transport and logistics: a systematic review of the literature. *International Journal of Production Research* 58(7): 2063–2081.

Saberi, S., M. Kouhizadeh, J. Sarkis, and L. Shen. 2019. Blockchain technology and its relationships to sustainable supply chain management. *International Journal of Production Research* 57(7): 2117–2135.

Sangari, M. S., and J. Razmi. 2015. Business intelligence competence, agile capabilities, and agile performance in supply chain: An empirical study. *The International Journal of Logistics Management.*

Sarmah, S. S. 2018. Understanding blockchain technology. *Computer Science and Engineering* 8(2): 23–29.

Sayogo, D. S., J. Zhang, L. Luna-Reyes, H. Jarman, G.Tayi, D. L. Andersen, T. A. Pardo, and D. F. Andersen. 2015. Challenges and requirements for developing data architecture supporting integration of sustainable supply chains. *Information Technology and Management* 16(1): 5–18.

SCM Trends. 2019. Blockchain helps Nestlé to improve transparency in the food chain. https://www.supplychainmovement.com/blockchain-for-improving-transparency-in-the-food-chain/, access date 17th June 2022.

Scott, B., J. Loonam, and V. Kumar. 2017. Exploring the rise of blockchain technology: Towards distributed collaborative organizations. *Strategic Change* 26(5): 423–428.

Shaharudin, M. R., K. Govindan, S. Zailani, K. C. Tan, and M. Iranmanesh. 2017. Product return management: Linking product returns, closed-loop supply chain activities and the effectiveness of the reverse supply chains. *Journal of Cleaner Production* 149: 1144–1156.

Sristy, A. 2021. Blockchain in the food supply chain - What does the future look like?" https://one.walmart.com/content/globaltechindia/en_in/Tech-insights/blog/Blockchain-in-the-food-supply-chain.html.

Swan, M. 2015. *Blockchain: Blueprint for a new economy*. O'Reilly Media, Inc.

Tian, F. An agri-food supply chain traceability system for China based on RFID & blockchain technology. In *2016 13th International Conference on Service Systems and Service Management (ICSSSM)*, pp. 1–6. IEEE.

Treleaven, P., R. G. Brown, and D. Yang. 2017. Blockchain technology in finance. *Computer* 50(9): 14–17.

Tseng, M. L., M. Lim, and W. P. Wong. 2015. Sustainable supply chain management: A closed-loop network hierarchical approach. *Industrial Management & Data Systems* 115(3): 436–461.

Uniliver News. 2019. Explainer: what is blockchain, who's using it and why? https://www.unilever.com/news/news-search/2019/explainer-what-is-blockchain-whose-using-it-and-why/, access date 17th June 2022.

Vrchota, J., M. Pech, L. Rolínek, and J. Bednář. 2020. Sustainability outcomes of green processes in relation to industry 4.0 in manufacturing: systematic review. *Sustainability* 12(15): 5968.

Wang, Q., and M. Su. 2020. Integrating blockchain technology into the energy sector—from theory of blockchain to research and application of energy blockchain. *Computer Science Review* 37: 100275.

Wang, S., and B. R. Sarker. 2004. A single-stage supply chain system controlled by kanban under just-in-time philosophy. *Journal of the Operational Research Society* 55(5): 485–494.

Yaga, D., P. Mell, N. Roby, and K. Scarfone. 2019. Blockchain technology overview. *arXiv preprint arXiv:1906.11078*.

Yli-Huumo, J., D. Ko, S. Choi, S. Park, and K. Smolander. 2016. Where is current research on blockchain technology?—a systematic review. *PloS One* 11(10): e0163477.

Zheng, Z., S. Xie, H. Dai, X. Chen, and H. Wang. 2017. An overview of blockchain technology: Architecture, consensus, and future trends. *In 2017 IEEE international congress on big data (BigData congress)*, pp. 557–564. IEEE.

Section II
Data Mining, Computational Framework, and Practices

CHAPTER 7

Mathematical Model of Consensus and its Adaptation to Achievement Consensus in Social Groups

Iosif Z Aronov[1,]* and *Olga V Maksimova*[2]

Contents

[1] Prof., Doctor of Technical Sciences, MGIMO (Moscow State Institute of International Relations), University of the Ministry of Foreign Affairs of the Russian Federation.
[2] Candidate of Technical Sciences, MISiS (Moscow Institute of Steel and Alloys), YU. A. IZRAEL INSTITUTE OF GLOBAL CLIMATE AND ECOLOGY.
* Corresponding author: aronoviz48@gmail.com

7.1 Introduction

Consensus as a method of decision-making is often used in the activities of social groups. In the context of this research, by a social group, we mean a set of people of a fixed number that interact with each other to achieve a specific goal. In this regard, the passengers of an aircraft are not a social group, and negotiators can be considered as a social group. Currently, in many international, regional, and national technical committees for standardization, international organizations (for example, OSCE, IPCC, WTO), European communities, and social movements, decision-making is based on the consensus of the group members. In voting, the majority wins over the minority. Consensus allows for finding a solution that each member of the group supports or, at least, considers acceptable. This approach ensures that all the opinions of the group members, their ideas and needs will be considered.

In this study, we consider consensus as the absence of fundamental disagreement among the members of a social group regarding the issue under consideration.

The main rule of consensus-based decision-making (in short consensus) is currently widely used in various dialogues (bilateral or multilateral negotiations), in the work of international organizations (e.g., IEC, ISO, APEC (Asia-Pacific Economic Cooperation)), in the activities of parties and political organizations, and for the development of medical recommendations.

Comparing the consensus as a method of decision-making with voting, it should be noted that voting initially generates competition (and not cooperation), and does not consider the possibility of compromise because the minority submits to the majority opinion (which does not happen, because the minority, as a rule, remains unconvinced), and disturbs cohesion of a society or a group (Della Porta et al. 2005). The problems of reaching a consensus in a social group, which is based, as a rule, on the ability and capacity of its members to find a compromise, have been poorly investigated.

The main problem of studying the achievement of consensus in social groups is related to the complexity of organizing such a study, especially for large groups, with more than five participants. Therefore, it is advisable to investigate the phenomenon of consensus in large social groups using modeling methodology.

It is important to note that the study of consensus in practice within the framework of social psychology raises many questions related to ensuring the reproducibility of research results. As follows from the fundamental study of the collaboration of evidence-based psychologists (Open science collaboration 2015), out of 100 original experimental studies in the field of social psychology, other groups managed to reproduce no more than

39 experiments. A "good" mathematical consensus model using simulation is free of this systemic defect.

The paper of DeGroot (DeGroot 1974) demonstrated the principal possibility of describing the process of achieving consensus based on regular Markov chains (Gantmacher 1959). Recently, this model has been applied in different fields, for example, in network automation management (Chebotarev 2010), negotiation process (Mazalov and Tokareva 2012), and multiagent behavior (Kozyakin et al. 2019).

In practice, it is important to assess the speed of the convergence of experts' opinions concerning the parameters of various social groups: the number of members, their authoritarianism[1] and the magnitude of discrepancies in the opinions of members of the group, which has not been analyzed so far.

Also, this research intends to consider various details associated with building consensus in conditions of coalition formation in social groups. Coalition formation is a dynamic process in any social group (Myers 2010). Therefore, the issue of assessing the time of convergence of expert opinions, depending on various parameters that characterize the behaviour of members during the negotiation process in the presence of coalitions, is relevant. The mathematical model is constructed to achieve consensus in terms of coalitions, which are overcome during the negotiation process by concessions. It was shown that in some situations the number of iterations (agreements, negotiations) is very large. Moreover, in the decision-making process, there is always a risk of blocking a decision by a minority in the group, which not only prolongs the decision-making time but even makes it impossible.

As a rule, such a minority is presented by one or two odious people. Such a member of the group tries to dominate the discussion, and always stands by his/her opinion, ignoring the position of the others (Sheril et al. 2012).

To overcome this problem, it was proposed to make decisions based on the principle of "Consensus Minus One" or "Consensus Minus Two". This position does not take into account the opinion of one or two odious members of the group. It is not for nothing that one of the consensus theorists in the social movement P. Gelderloos (Gelderloos 2006) called such people "weeds". Such a "weed" can be a dominant member of the group who ignores the opinions of others.

For example, in standardization, this approach has led to the formation of a new type of standardization document—incomplete consensus

---

[1] **Authoritarianism** (from the Latin *autoritas* – influence, power) – social and psychological feature of an individual that reflects the desire to maximally submit the interaction and communication partners.

documents. The effectiveness of these documents was manifested during the COVID-19 pandemic when many national standardization bodies in order to accelerate the process began to develop standardization documents on personal protective equipment (for example, face masks) based on incomplete consensus standards instead of common consensus standards (Aronov et al. 2021).

Obviously, the "Consensus Minus One" or "Consensus Minus Two" rule reduces the decision-making time compared to the classical consensus principle, but by how much?

To solve this problem, the authors undertook a corresponding study based on the mathematical method of Markov chains, which was described in a paper of scholars (Aronov et al. 2018) and which proved its suitability for simulations.

7.2 Description of the Model for Consensus Based on Regular Markov Chains

In the paper (Aronov et al. 2018) a theoretical model for achieving consensus based on regular Markov chains is described. Here are the main results. Let n be the number of group members participating in the discussion; let $S(0) = (s_{01};...; s_{0n})$ is the vector of the initial opinions of the group members, where s_{0i} is the opinion of the i-th member. The participants in the negotiation process exchange opinions among themselves regarding the values of the S vector. The opinion of each of them may change during the negotiations. Introducing the probability of trust of the i-th participant to the opinion of the j-th participant through $0 < p_{ij} < 1$ ($i = 1, ..., n; j = 1, ..., n$) a square trust matrix $P = (p_{ij})$, is formed, which sets the sequential process of agreeing the views of group members. The sum of the probabilities in each row of the matrix is 1, i.e., for any $i \in (\overline{1,n})$ the following $\sum_{j=1}^{n} p_{ij} = 1$ is true. The vector of opinions of the group members at each step of the negotiations can be calculated by the formula

$$S^T(1) = P \cdot S^T(0) = (s_{01},...,s_{0n})^T \tag{1}$$

After the k-th step of negotiations, the vector of opinions can be calculated by the formula

$$S^T(k) = (s_{k1},...,s_{kn})^T = P \cdot S^T(k-1) = P^k \cdot S^T(0) \tag{2}$$

The iteration process ends at the m-th step if all the rows of the P^m matrix become the same. Thus, the trust matrix P after m iterations reaches the final F matrix (Kemeny and Snell 1960). Mathematically, this means that the confidence matrix P after m iterations has reached the final matrix F, and since the final matrix F does not change with subsequent iterations, the expert opinion vector $S^T(m) = P^m \cdot S^T(0) = (s_m^{~1}, \ldots, s_m^{~n})^T$ does not change accordingly. This agrees with the well-known concept of the group dynamics theory describing the processes that take place in social groups (Myers 2010).

The necessary and sufficient condition for the convergence of the initial matrix P to the final matrix F (the necessary and sufficient condition for reaching consensus) for any vector of initial opinions is the regularity[2] of the matrix P (theory of Markov chains: Gantmacher 1959). In other words, it is necessary and sufficient that the sum over the rows of the matrix P is equal to 1 and that for any probabilities p_{ij} the strict inequality $0 < p_{ij} < 1$ is fulfilled. As it was written above, in terms of social group activity it is important that the members are loyal.

Thus, if the matrix of trust P is regular (that is, there are no ambitious experts with a distinct opinion), then whatever the initial views of the group members were, a consensus is achievable, although it may require a significant number of iterations (discussions in the social group). This means that in some cases, even with loyal experts, considerable time is required to achieve it.

7.3 Specific Cases in The Model of Attaining Consensus

The fact that the considered model is workable is demonstrated by the analysis of situations with violations of the expert loyalty condition (Zazhigalkin et al. 2014).

7.3.1 Coalitions

This is another case of the impossibility of a consensus, which is associated with the formation of coalitions within the group. The corresponding matrices are decomposable. In this situation a consensus cannot be reached for any $n > 2$. In the literature on group dynamics, similar situations are considered, and conclusions are given (Gençer 2019). The task of the group leader is to eliminate the existing group coalitions through the choice of compromise solutions.

7.3.2 Domination

If there is one authoritative expert ($\exists i = \overline{1,n}: p_{ii} = 1$) in the group, then his opinion as a result of the approvals (iterations) does not change (in the final matrix F, the element p_{ii} remains equal to 1). The presence of both an

---

[2] The matrices, for which the sums of elements of all rows are equal to 1, are called stochastic. If for some n all elements of the matrix P^n are not equal to 0, then such a transition matrix is called regular.

authoritative member and a member with a level of self-confidence close to 1 delays consensus for a long time. Indeed, such a member of the group is difficult to convince. Therefore, the inclusion of an ambitious member should be strongly checked since the opinion of this particular expert will prevail.

7.3.3 Presence of One or Several Autocratic Group Members

The situation when there are several leaders in a group is characterized by a matrix $\mathbf{P}$ in which several numbers on the main diagonal are equal to 1. Matrices of this type and the Markov chains corresponding to them are said to be decomposable (Zazhigalkin et al. 2019).[3] In this situation, a consensus is never achievable for any $n > 2$. In the literature on group dynamics, similar conclusions are made (Gençer 2019). The presence of several autocratic members in a group fundamentally distinguishes this situation from the previous one. The presence of one autocratic member in a group provides a consensus which would be of low quality however in terms of the required number of approvals, whereas the presence of several autocratic members in a group leads to a total impossibility of consensus. This was confirmed by a sufficiently large number of observations obtained during monitoring of various groups' work: the more ambitious members, the more difficult it is to attain consensus in the group (Gençer 2019).

7.3.4 Global Domination

If in the group all experts have high self-esteem (i.e., we can assume that $\forall\ i = \overline{1,n}$: $p_{ii} = 1$, then the trust matrix $\mathbf{P}$, in this case, becomes a matrix of 1's. Since for any number of m iterations $\mathbf{P}^m = \mathbf{E}^m = \mathbf{E}$, the matrix $\mathbf{P}$ does not converge to the final one, and consequently, the consensus in this case in principle is unattainable.

7.3.5 Responsibility Shift

The situation in which each member shifts full responsibility to another member of the group, escaping responsibility for making a decision: members join the opinion of the group, considering it to be correct and their assessment to be erroneous. In the theory of Markov chains, it is shown that the corresponding transition matrix does not converge to the final matrix (Gantmacher 1959). Consequently, for such a group, a

---

[3] A matrix A is said to be decomposable if by rearrangements of rows it can be reduced to the form $\tilde{A} = \begin{pmatrix} B & 0 \\ C & D \end{pmatrix}$, where B and D are square matrices.

consensus is not achievable. In fact, at least two "irresponsible" experts will be enough not to reach a consensus in a group. The analysis of group dynamics shows that this situation is very common in life.

7.4 Analysis of The General Case in The Consensus Model

Let us evaluate the impact of the number of group members and their authoritarianism on the time of consensus-building using statistical modeling. The influence of the number n of group members on the time to reach consensus was investigated using statistical modeling. As a dependent variable, m is the number of discussions in the group before consensus is reached, at which the condition is met.

$$\det \mathbf{P}^m < \varepsilon. \tag{3}$$

The modeling involved several stages:

The first stage is the choice of factor's levels (the number of group members, n) for practical reasons: 5; 10; 20; 30; 40; 50. It seems that the chosen boundaries of the modeling parameters correctly describe the actual situation in groups.

The second stage is the selection of factor 2 levels (probabilities p_{ii}, which show the member's self-confidence): 0.20–0.30; 0.45–0.55; 0.65–0.75; 0.85–0.95; 0.90–1.00. In reality, the group members combine the behavioral implications of all groups (from authoritarian to conformist), which is reflected in the simulation conditions $0.2 < p_{ii} < 1$.

In the third stage—for each level n, the diagonal elements p_{ii} of the matrix $\mathbf{P}$ were modeled according to the uniform distribution law; the parameters of the uniform law were the boundaries of the corresponding level of the factor p_{ii}. In each row of the matrix $\mathbf{P}$, the residual probabilities $p_{ij} (i \neq j)$ were also modeled according to a uniform law with parameters 0 and 1 and normalized so that the sum of the probabilities within each row was equal to 1, that is, the matrix $\mathbf{P}$ became stochastic.

The accuracy of the matrix elements modeling was 0.01, which was derived from the considered boundaries of the number n of members in the group. The values of ε in the inequality (Equation 3) were determined by the accuracy of modeling of the elements in the matrix $\mathbf{P}$ and its size n:

$$\varepsilon = \left(\frac{0.01}{n-1} \right)^n, n \geq 2.$$

This type of dependence of ε on the number of members n was derived from:

(1) the change of accuracy of $(n-1)$ element in each row of the matrix under normalization.

(2) the power-law dependence related to the technique of calculating the determinant of the n-th order.

To obtain consistent conclusions on average number of approvals m while changing other parameters, 100 simulations were carried out in Excel at each fixed level of factors n and p_{ii} (Efron and Tibshirani 1991). We analyzed both the impact of authoritarianism on the group members and their number on the number of meetings necessary to reach a consensus. A three-dimensional model of the dependence is shown in Fig. 1, and two-dimensional dependencies are presented in Figs. 2 and 3.

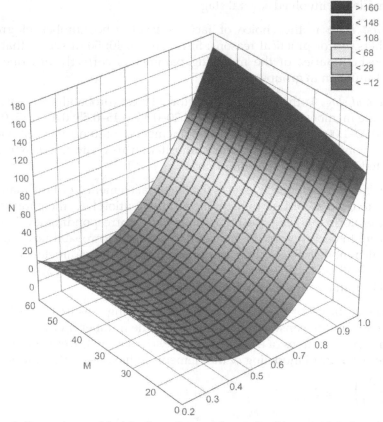

Figure 1: Regression model of the dependence of the number of approvals before reaching consensus on the number of members (the left axis) and the level of authoritarianism of members (the right axis).

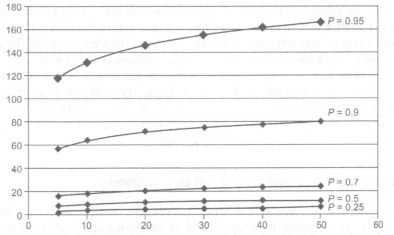

Figure 2: Dependence of the number of approvals m (axis y) on the number of group members n (axis x): the corresponding average values of the authoritarian level **P** = are shown next to the curves.

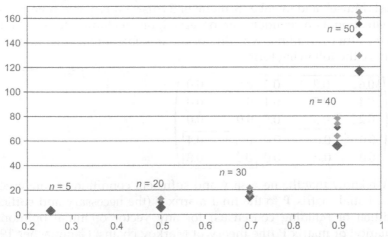

Figure 3: Dependence of the number of approvals m (axis y) on authoritarianism (axis x) under a fixed number of group members n.

It was found that the most obvious was the relationship between the number of approvals and the level of authoritarianism p_{ii} of the group members for a fixed number of members (Fig. 3). The graphs clearly illustrate the high sensitivity of the number of approvals m to the growth of authoritarianism, especially stating from p_{ii} equal to approximately 0.8.

With an increase in the number of committee members, a gradual power-law rise of the number of approvals was observed (Figure 2). The analysis of the model confirmed not only the visual but also a theoretically good agreement with the data (for each curve $R^2 = 0.97$). The study shows that the increase in the number of members of the group (with a fixed level of authoritarianism of members) leads to an increase in the number of iterations m to reach consensus (Figure 3). At the same time, in the case of a low level of authoritarianism (< 0.5) its influence decreases.

Another fact attracts the attention: with the increase in the number of members of the group and the increase in the authoritarianism of its members, the spread of the numbers of approvals m to attain consensus (relative to average values) on average increases, which indicates the growth of the disunity in the expert group.

7.5 Consensus Building Model in Coalitions

The formation of coalitions is a natural process due to the formation of the interests of different groups. Let's illustrate the mathematical model with two coalitions for a group of five members: member 1 trusts himself and members 2 and 3; member 2 trusts himself and members 1 and 3; respectively, member 3 trusts himself and members 1 and 2; group member 4 trusts himself and member 5; member 5 trusts himself and member 4. In this situation, two coalitions are observed, one including three members and the other including two. One of the possible initial trust matrices **P** can have the following form:

$$
P = \begin{pmatrix}
0.6 & 0.2 & 0.2 & 0.0 & 0.0 \\
0.4 & 0.5 & 0.1 & 0.0 & 0.0 \\
0.2 & 0.3 & 0.5 & 0.0 & 0.0 \\
0.0 & 0.0 & 0.0 & 0.6 & 0.4 \\
0.0 & 0.0 & 0.0 & 0.2 & 0.8
\end{pmatrix}
$$

We know that the necessary and sufficient condition of convergence of the initial matrix **P** to the final matrix **F** (the necessary and sufficient condition of building consensus) for any vector of initial opinions is regularity[4] of matrix **P** (the theory of Markov chains: Gantmacher 1959). In other words, it is necessary and sufficient that sums of rows of matrix **P** are equal to 1 and the inequality $0 < p_{ij} < 1$ is satisfied for any rows for probabilities p_{ij}. In terms of the discussion of issues in a social group, it is important that some of its members have their own opinions and have confidence in the opinions of individual colleagues. The regularity condition is violated in the coalition model, so the consensus is unattainable in this case. Matrices of this kind and their corresponding Markov chains are decomposable.

---

[4] Matrices, which sums of elements of all rows are equal to one, are called stochastic. If for some n all elements of the matrix P^n are not equal to zero, then such transition matrix is called a regular matrix.

The established coalitions can be eliminated by choosing compromise solutions. These solutions can be provided in different ways. Let's consider a situation in which there is a redistribution of trust probabilities p_{ij} of the i expert of one coalition to the j expert of the second coalition, which is achieved by additional reasoning in the process of discussion. Such a situation (redistribution of trust probabilities) will be called concessions.

This corresponds to the model outlined in the work of social psychologist Homans (1958). It proposes a scheme of functioning of small social groups on the basis of the exchange of utilities.

Theoretically, we can talk about two options for concessions:

(1) unilateral concessions when members of one coalition redistribute their trust probabilities to members of a second coalition.

(2) ambilateral concessions in which members of one coalition redistribute their trust probabilities to members of the second coalition and, accordingly, members of the second coalition redistribute their probabilities to members of the first coalition.

Below there is a discussion about the analysis of the time required to build a consensus on unilateral concessions.

Unilateral concessions issues in the negotiation process have been addressed in several works, for example, in (Seungwoo et al. 2004). However, they did not analyze the factors influencing the time of the negotiation process. Note that the article (Kudish et al. 2015) concluded that the uniqueness of the conflict affects its time. This aspect is not considered in the present article.

7.5.1 Unilateral Concession by Members of a Small Coalition

Let's build a model of unilateral concessions of two coalitions. Consider a group of 20 members. It was shown that this number of experts is optimal in terms of the number of negotiations before building a consensus under other equal conditions.

Let's assess the convergence time of the opinion matrix $\mathbf{P}$ to the final matrix $\mathbf{F} = \mathbf{P}^m$. This time is determined by the required number of m iterations (discussions) to form a consensus. The value m will be calculated by matching all the corresponding column elements with two decimal places.

For simulation, we have the following conditions: number of experts $n = 20$, m—the number of meetings before consensus is built (i.e., the time for consensus building), and two coalitions. Let's introduce an index of coalition influence I, equal to the ratio of the number of members of the larger coalition to the smaller one. In practice, this can be interpreted as the quantitative "strength" of one coalition in relation to the other.

Simulation for the case of unilateral concessions by members of a small coalition consisted of several stages.

(1) *First stage.* The levels of change in the coalition's influence index I, were chosen:

1st level: $I = 4$ (coalitions with 16 and 4 members);

2nd level: $I = 1.5$ (coalitions with 12 and 8 members);

3rd level: $I = 1$ (in coalitions of 10 members each).

(2) *Second stage.* Levels of concessions among members of a small coalition were selected. Numerous studies show that in small groups group opinion is especially important, they are more consolidated. Thus, personal contacts allow for all members to participate in the development of group opinion and control of conformity of group members in relation to this opinion (Gockel et al. 2008, Kerr 1989). Therefore, members of small groups are ready to consciously concessions equal in relation to their members for the further conduct of negotiations. The following levels are distinguished for concessions of Y: $Y = 10\%$; $Y = 20\%$; $Y = 33\%$; $Y = 50\%$; $Y = 75\%$.

Each level specifies a percentage value for the small coalition, making concessions that are evenly distributed among the members of the other coalition. Consider an example of a transformation of the trust matrix $\mathbf{P}$ for five experts with two coalitions after concessions by members of the smaller coalition (3 members in the larger coalition and 2 members in the smaller coalition):

$$
\mathbf{P} = \begin{pmatrix}
0.6 & 0.2 & 0.2 & 0.0 & 0.0 \\
0.4 & 0.5 & 0.1 & 0.0 & 0.0 \\
0.2 & 0.3 & 0.5 & 0.0 & 0.0 \\
0.0 & 0.0 & 0.0 & 0.5 & 0.5 \\
0.0 & 0.0 & 0.0 & 0.2 & 0.8
\end{pmatrix} \Rightarrow \tag{4}
$$

$$
\tilde{\mathbf{P}} = \begin{pmatrix}
0.6 & 0.2 & 0.2 & 0.0 & 0.0 \\
0.4 & 0.5 & 0.1 & 0.0 & 0.0 \\
0.2 & 0.3 & 0.5 & 0.0 & 0.0 \\
0.2 & 0.2 & 0.2 & 0.2 & 0.2 \\
0.2 & 0.2 & 0.20 & 0.8 & 0.32
\end{pmatrix}
$$

Concession means that a coalition of two members "concedes" a probability of 0.6, which is uniformly "passed on" to members of the larger coalition.

We will count the number of concessions m from this point onward. For the considered matrix $\tilde{P}$ with number $m = 6$ the final matrix F with the accuracy of its elements $\varepsilon = 0.01$ will have the form

$$F = \begin{pmatrix} 0.44 & 0.32 & 0.24 & 0.00 & 0.00 \\ 0.44 & 0.32 & 0.24 & 0.00 & 0.00 \\ 0.44 & 0.32 & 0.24 & 0.00 & 0.00 \\ 0.44 & 0.32 & 0.24 & 0.00 & 0.00 \\ 0.44 & 0.32 & 0.24 & 0.00 & 0.00 \end{pmatrix} \tag{5}$$

Value $\varepsilon = 0.01$ for each j-th column of this F-matrix (5) determined the following absolute difference $|p_{ij} - p_{kj}| < \varepsilon$ for all rows i,k ($j,i,k = \overline{1,5}$). Formula (3) makes it possible to obtain an estimate of the accuracy only in the general case. In the case of two coalitions and the presence of zero probabilities, it becomes insufficient.

Importantly, in the case of unilateral concessions, the consensus decision takes into account only the opinion of the coalition that accepts the concessions. Mathematically, this follows from the properties of matrices of the form (4) (DeGroot 1974), but in practice, it means that concessions reduce a party's own demands in order to allow the other party to achieve the desired result.

(3) At the *third stage* for each level n modeling of elements p_{ij} of matrix P was carried out using uniform law of distribution under given conditions of the first and second stages so that the sum of probabilities within each row equals 1, i.e., so that matrix P became stochastic. In order to obtain stable conclusions about the average number of m variation of other parameters, 100 simulations were performed on each fixed level of factors I and Y in Excel (Efron and Tibshirani 1991). The number was defined as the degree of matrix P, at which in the final matrix F the elements within each column satisfied the condition $|p_{ij} - p_{kj}| < \varepsilon$ for all i,k ($j,i,k = \overline{1,20}$).

Figure 4 shows the dependence of the number of iterations (negotiations) m on the influence index of the coalition I at fixed levels of concessions Y. It turns out that at significance level $\alpha = 5\%$ the average values of number of negotiations at the level of concessions greater than 10% are statistically indistinguishable for different influence indices I. It can be concluded that the size of a large coalition has little effect on the number of negotiations before consensus is built.

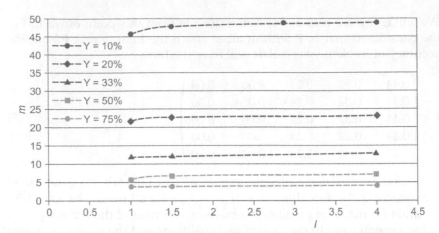

Figure 4: Dependence of the average number of negotiations m on the coalition influence index I at fixed levels of unilateral concessions Y by members of a small coalition.

For the average number of negotiations among different influence indices I of coalitions a suitable regression is drawn, which has the form (Tukey 1981).

$$\hat{m} = \hat{a} \cdot Y^{\hat{b}}, \tag{6}$$

where $\hat{m}$ is the regression value of the number of iterations to build a consensus, Y is the numerical value of concessions, $\hat{a}$, $\hat{b}$ are the equations factors.

There is a gradual drop in the number of concessions with increasing value of concessions. Analysis of this model (2) confirmed not only visual but also theoretical good negotiation with model data (for each curve $R^2 \approx 0.997$, Figure 5). For example (4) at $Y = 60\%$, the average number of negotiations m according to the calculated model is $m = 879 \cdot 60^{-1.2} \approx 6$, which coincides with the experimentally obtained early value. The diagrams in Figs. 4 and 5 illustrate the high sensitivity of the m number to the level of concessions Y among the members of the small coalition, and the weak sensitivity for the index of coalition influence I.

Consensus is always achievable for the members of a small coalition. In this case, the lesser the concessions, the more time it takes to build a consensus. Even in small concessions, i.e., dedication to compromise leads to a consensus. An increase in the number of concessions leads (among all other equal factors) to a sharp decrease in time before consensus (Aronov and Maksimova 2020).

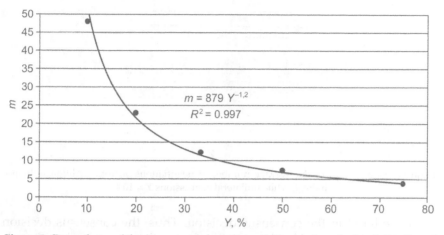

$$m = 879\ Y^{-1,2}$$
$$R^2 = 0.997$$

Figure 5: Dependence of the average number of negotiations m from the level of unilateral concessions Y by members of the small coalition (index of coalition influence $I = 4$).

7.5.2 Unilateral Concessions by Members of a Large Coalition

Early it is found that the influence of the index I on the number of negotiations before the consensus is built is insignificant when the value of unilateral concessions is greater than 10%. Consider the unilateral concessions model $Y = 10\%$ from the side of a large coalition and compare with the results of Section 1. In order to obtain stable conclusions about the average number of m concessions at each fixed level of the coalition's influence index, 100 simulations were performed in the Excel environment. The results presented in Fig. 6 show that for small concessions $Y = 10\%$ the fact of which coalition makes it—small or large—is not important. In general, the more differences are observed in the number of coalitions, the greater the difference in the average number m before building a consensus (Aronov and Maksimova 2020). However, these differences are statistically insignificant at the level of $\alpha = 5\%$ for all coalitions. At any index of coalition influence, the significance of "strength" coalitions for convergence is levelled.

In social groups this is confirmed by the way coalitions interact:

(1) in the case of concessions on the part of a small group, a large group "pulls over" the opinion of a small group by its "strength" (number of members), i.e., there is a quantitative nature of the process of negotiations.

(2) when concessions are made by a large coalition, the reason for concessions can only be a strong argument for the position of the small group, because the opinion of the large coalition is not

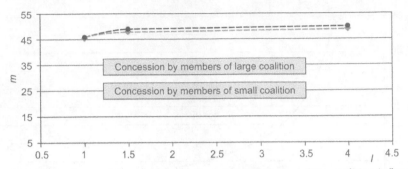

Figure 6: Dependence of the average number of negotiations m from coalition influence index I, while unilateral concessions $Y = 10\%$.

considered in the consensus decision. Thus, the consensus decision will be determined by the opinion of a small group, meaning, there is a qualitative character to negotiations.

In each of these methods, the equal average number of negotiations before the consensus is built is provided by the different nature of the interaction between coalitions in the group. Authors (Gockel et al. 2008, Kerr 1989) in their papers came to the following conclusions:

(1) small groups are more cooperative, rather than larger groups, and make better use of resources, available to them.

(2) one of the explanations for the difficulties of large groups is that, as the number of group members increases, each person's behavior becomes less identifiable, which is likely to lead to an increase in "stowaways". Large groups are more likely to have difficulties coordinating the efforts of individuals, and this may reduce the quality of cooperation.

Thus, it was revealed that even a small concession leads to the onset of consensus. Increasing the size of the assignment results (with other factors being equal) in a sharp decline in time before the consensus.

7.6 Presence of One or Several Autocratic Group Members Domination

The presence of an authoritarian expert (for example, $p_{11} = 1$) also describes the situation of two coalitions with a number of members in 1 and $(n-1)$—in the other. Consensus can be built only in conditions of trust for the authoritarian member (as follows from the paper by Aronov et al. 2018), however, for a rather significant number of iterations (negotiations), given that the authoritarian member (if not conceded) always remains with his opinion, meaning that the authoritarian member "pulls" other members towards his opinion. In other words, the quality of such a consensus is relatively low.

In the case when there is one autocratic participant ($\exists i = \overline{1;n}$, $p_{ii} = 1$), there is an absorbing state in the trust matrix **P**, and it can no longer leave the iteration process (Kemeny and Snell 1960). Within the framework of negotiations, this means that the opinion of such a participant does not change because of agreements/iterations (in the final matrix **F**, it is p_{ii} element that remains equal to one). It is difficult to convince such a member of the group, a consensus can be achieved only by taking into account the opinion of this autocratic leader. Let's consider an example of a possible initial trust matrix **P**:

$$\mathbf{P} = \begin{pmatrix} 0.6 & 0.1 & 0.1 & 0.1 & 0.1 \\ 0.4 & 0.1 & 0.2 & 0.2 & 0.1 \\ 0.1 & 0.3 & 0.1 & 0.3 & 0.2 \\ 0.3 & 0.1 & 0.1 & 0.1 & 0.4 \\ 0 & 0 & 0 & 0 & 1 \end{pmatrix} \tag{7}$$

As a result of the iteration process described above, the initial trust matrix **P** will converge to the final matrix **F** of the following form:

$$\mathbf{F} = \begin{pmatrix} 0 & 0 & 0 & 0 & 1 \\ 0 & 0 & 0 & 0 & 1 \\ 0 & 0 & 0 & 0 & 1 \\ 0 & 0 & 0 & 0 & 1 \\ 0 & 0 & 0 & 0 & 1 \end{pmatrix}.$$

In such a group, a consensus is achievable. For example, with the initial trust matrix (7), it will take $m = 29$ agreements to reach a consensus (under conditions of a given accuracy of the matrix elements $\varepsilon = 0.01$, defined as the absolute difference for each j-th column $|p_{ij} - p_{kj}| < \varepsilon$ for all rows i,k ($j,i,k = \overline{1,5}$), which for such a small group is a long process. It is important that in this case, the consensus considers only the opinion of the autocratic leader, the opinion of the other members of the group is not taken into account in the final decision. In other words, the quality of consensus also degrades. If there is another autocratic member in the group in example (7), we get the following **P** matrix:

$$\mathbf{P} = \begin{pmatrix} 0.6 & 0.1 & 0.1 & 0.1 & 0.1 \\ 0.4 & 0.1 & 0.2 & 0.2 & 0.1 \\ 0.1 & 0.3 & 0.1 & 0.3 & 0.2 \\ 0 & 0 & 0 & 1 & 0 \\ 0 & 0 & 0 & 0 & 1 \end{pmatrix}.$$

Such a matrix already contains two absorbing states (Kemeny and Snell 1960). Matrices of this type and their corresponding Markov chains are expanding (Gantmacher 1959). Since the product (and, accordingly, the degree) of expanding matrices is an expanding matrix, it is obvious that in this situation consensus is not achievable (for any $n > 2$). In the literature on group dynamics, similar conclusions are made when it comes to the presence of several autocratic group members (Myers 2010). Therefore, for further research of decision-making based on the "Consensus Minus Two" rule, we will consider a model with one autocratic member (leader) and a second member with a high level of authoritarianism, which is though not equal to one.

It is possible to solve the current situation with the protraction of negotiations by their reorganization, which can be performed in various ways:

(1) replacement of autocratic members.

(2) ignoring the opinions of autocratic members in the group when deciding.

The second approach is more common (we will call it "consensus minus k") since it does not require additional resources to find a new member of the group and gives a significant reduction in the number of agreements. For the trust matrix (7), after the removal of the autocratic leader, the number of agreements in the group is reduced to $m = 4$, i.e., it is reduced by 86% compared to the initial $m = 29$. Therefore, the presence of even one ambitious member in the group should be suppressed, since the opinion of this participant will prevail, and the negotiation process will be protracted.

7.6.1 Time Required to Reach a Consensus When There are One or Two Autocratic Group Members

Let us to build a mathematical model of dominance using the example of 20 members of a social group. The paper by Zazhigalkin, Aronov, and Maksimova (2019) testifies to the fact that this number of members is optimally appropriate in terms of the number of agreements before a consensus, provided that all other conditions are the same.

Let us estimate the time of the opinion matrix $\mathbf{P}$ required for it to converge to the final matrix $\mathbf{F} = \mathbf{P}^m$. This time is determined by the necessary number of iterations m (discussions among social group members) to build a consensus. Mathematically, m is defined as the degree of the matrix $\mathbf{P}$, at which the elements within each column j of the final matrix $\mathbf{F}$ meet the common condition $|p_{ij} - p_{kj}| < \varepsilon$ for all i,k ($j,i,k = \overline{1,20}$). To calculate the m value, we will refer to the condition $\varepsilon = 0.01$.

So, we consider two cases of simulation:

(1) $n = 20$, where n stands for the "number of social group members", m stands for the "number of agreements before a consensus" (i.e., the time required to reach a consensus), and there is one leader running the group.

(2) $n = 20$, where n stands for the "number of social group members", m stands for the "number of agreements before a consensus", there is one leader running the group and also one profoundly autocratic member, since one more member who shows the ultimate level of autocracy would prevent it from reaching a consensus.

Each simulation process consisted of several stages.

(1) At the *first stage*, we chose the levels of change in the number of group members (n): $n = 5$; $n = 10$; $n = 20$.

(2) At the *second stage*, we suggested the probability p_{ii} to set the likelihood of the participant's confidence in themselves (so-called "the level of authoritarianism"). When it comes to the leader, p_{nn} is 1; we use for profoundly autocratic members $p_{n-1,n-1} = 0.85 \div 0.95$. For other group members, please see the levels of authoritarianism p_{ii} presented below:

1st level: $p_{ii} = 0.20 \div 0.30$;

2nd level: $p_{ii} = 0.45 \div 0.55$;

3rd level: $p_{ii} = 0.65 \div 0.75$;

4th level: $p_{ii} = 0.85 \div 0.95$.

If the level of authoritarianism p_{ii} reaches almost 1, the person concerned is unlikely to be prone to compromise, and if the p_{ii} value is around 0, is typical of a conformist who does show a volatile position while negotiating, and, therefore, they are inclined to drift toward other people's opinions rather than to their own ones. Behavioral patterns of different group members were reflected in the conditions for simulation $0.20 \leq p_{ii} \leq 0.95$.

(3) At the *third stage*, we simulated the p_{ii} elements of the matrix **P** for each level n. For this purpose, we used the uniform law of distribution law under the given conditions so that the sum of the probabilities within each line did equal 1. In order to draw robust conclusions about the average number of agreements m against the backdrop of other parameters changing, there were 100 simulations conducted in the Excel environment for each level of factors (Efron and Tibshirani 1991).

We have already pointed out that the consensus will consider only the leader's opinion, if there is one in the group. Therefore, we will further investigate the relationship between the number of agreements and the number of members whose opinion is not considered in the resulting decision. Figure 7 shows the dependence of the number of agreements (iterations) m on the number of group members excluding k as autocratic group members when $k = 1;2$. We may conclude that the number of group members affects the number of agreements before a consensus strongly given that there are autocratic members present, which is consistent with the findings obtained previously (Zazhigalkin et al. 2019).

In order to present the average number of agreements on different levels of trust (please see the second stage of the simulation for details), we built appropriate linear regression dependencies, which have the following form (Tukey 1977).

$$\hat{m} = \hat{b} \cdot (n - k) + \hat{a}, \tag{8}$$

where $\hat{m}$ stands for the "regressive number of iterations required to reach a consensus", $(n - k)$ stands for the number of group members excluding autocratic ones (k), $\hat{b}, \hat{a}$ stand for regression coefficients of the equation. It stands to reason that the "Consensus Minus One" model, if there are no other members (i.e., $n - 1 = 0$), does not provide any agreements in contrast to the "Consensus Minus Two" model implying that, in presence of two autocratic members, the ultimately autocratic one must score several numbers of agreements to pull the other one to their side. Therefore, for

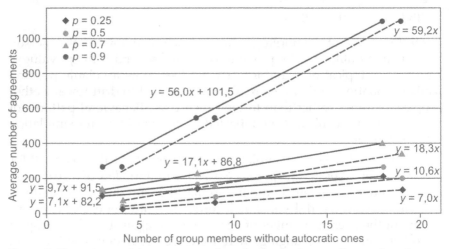

Figure 7: Dependence of the average number of agreements m on the number of group members for different levels of authoritarianism p: the "Consensus Minus One" model does not take into account one autocratic group member (linear trends are marked as "– –"), the "Consensus Minus One" model does not take into account two autocratic group members (linear trends are marked as "–").

the "Consensus Minus One" model, regression dependences contain a zero intercept ($\hat{a} = 0$, see Fig. 7). The analysis of the model attests not to only visual, but also theoretical good alignment with the simulation data (for each straight line, the coefficient of determination is $R^2 \approx 0.997$). It is evident that the number of agreements in the presence of two autocratic members is, on average, greater than in the presence of one, provided those other parameters are equal (Fig. 7).

Given that the model is of high quality, equation (8) allows deducing the average number of agreements per member for groups with one leader. This number is set by the coefficient $\hat{b}$; we may analyze its growth when the average authoritarianism of the group members changes. In terms of consensus, this number can be interpreted as a "specific" number of agreements. For example, if the authoritarianism of the group members is $p_{ii} = 0.2 \div 0.3$, the specific number of agreements equals 7. However, if $p_{ii} = 0.85 \div 0.95$ is the case, it increases to 59, i.e., almost 9 times (Fig. 7), which indicates a marked jump in the total number of agreements and a significant protraction in the negotiation process.

The graphs in Fig. 8 illustrate the high sensitivity of the number of agreements m to the average authoritarianism of the members p and reveal a hyperbolic connection when the average authoritarianism p approaches 1. This suggests a possible sharp increase in the number of agreements when building a group with autocratic members.

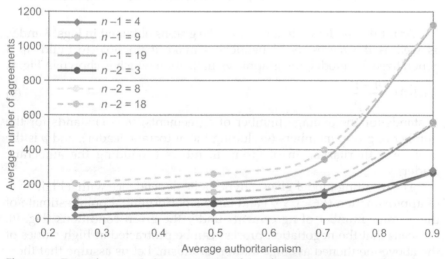

Figure 8: Dependences of the average number of agreements m on the average authoritarianism of group members for different numbers of members without taking into account one authoritarian for $(n - k)$ members when $k = 1;2$.

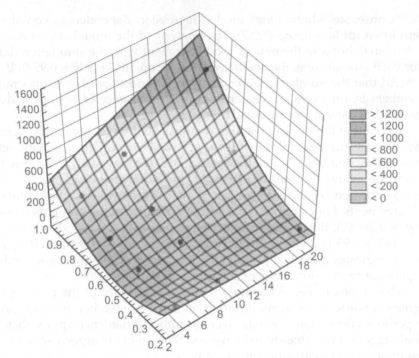

Figure 9: 3D imaging of the model that presents the dependencies of the average number of agreements on the number of group members (if there is one autocratic member).

As for the model with one leader, the graphs obtained in Figs. 8 and 9, as well as the regression dependences make it possible to build up a generalized 3D model, the graphic visualization which is shown in Fig. 9:

$$\hat{m} = 5.84 \cdot \frac{n-1}{1-p} \tag{9}$$

m stands for the average number of agreements, $(n - 1)$ stands for the number of group members (excluding the autocratic leader), and p is the average authoritarianism of group members (excluding the autocratic leader).

The resulting model gives a hyperbolic dependence with a high level of approximation ($R^2 \approx 0.997$) and allows performing a point estimate of the average number of agreements under the given conditions (Fig. 9). It means that the negotiation process can be protracted at high values of the above-mentioned average authoritarianism. Let us assume that there are 20 participants gathered to agree on the project, among whom there is one leader, and the average authoritarianism of the remaining members is $p = 0.9$. In this case, according to model (7), we might expect an average of 1041 agreements before a consensus, which gives evidence of an unacceptably long negotiation process of several years. One of the possible solutions, as described, is to remove the autocratic leader from the decision-making process.

Thus, in some situations, the number of iterations is very significant. Reaching a consensus requires considerable time since it is necessary to come to an agreement within the group, regardless of its size. Moreover, in the decision-making process, there is always a risk of blocking a decision by a minority in the group, which not only prolongs the decision-making time but even makes it impossible. As a rule, such a minority is presented by one or two odious people. Such a member of the group tries to dominate the discussion, always standing by his/her opinion, ignoring the position of the others. The decision-making process is delayed and the quality of consensus is deteriorating since only the opinion of the dominant part of the group would be taken into account. In order to decide how to solve this problem, it was decided to make a decision based on the principle of "Consensus Minus One" or "Consensus Minus Two", that is, not to take into account the opinion of one or two odious members of the group. For example, in climate research, where many scientific disciplines are involved, a complete consensus is almost impossible. The simulation results showed that its use can reduce the time required to reach a consensus to 97%, which is crucial for practice. It is shown that the negotiation process can be protracted at high values of the above-mentioned average authoritarianism.

7.6.2 Time Required to Reach a Consensus When There is a Leader in the Group

Consider a group in which there is one member who and only him (except himself) is trusted by all other members of the group, while this member of the group is not authoritarian, trusting the other members of the group. Then the confidence matrix is formed from the conditions:

$$\forall i = \overline{1;n} : 0 < p_{ii} < 1; \exists i_0 = \overline{1;n}\ \forall i \neq i_0, j \neq i_0 : p_{ij} = 0.$$

We will interpret this case as leadership (one participant is trusted by all members of the group, and he, in turn, trusts them).

Let us analyze how the structure of the consensus decision will change if the dominant character in this group is replaced by a leader who is trusted by the members of the group, and at the same time, he trusts the members of the group (the last row of the matrix (7)). The **P** matrix characterizes not only the mutual influence of the group members on each other but also the mutual influence of the leader on the group members and vice versa. This is consistent with the opinion of the sociologist G. Homans that the right to influence others is acquired at the cost of allowing others to influence themselves (Homans 1958).

Consider, for example, the following confidence matrix:

$$\mathbf{P} = \begin{pmatrix} 0.5 & 0 & 0 & 0 & 0.5 \\ 0 & 0.4 & 0 & 0 & 061 \\ 0 & 0 & 0.3 & 0 & 0.7 \\ 0 & 0 & 0 & 0.2 & 0.8 \\ 0.2 & 0.2 & 0.2 & 0.2 & 0.2 \end{pmatrix}.$$

In this case the final matrix **F** will look like

$$\mathbf{F} = \begin{pmatrix} 0.18 & 0.15 & 0.12 & 0.11 & 0.44 \\ 0.18 & 0.15 & 0.12 & 0.11 & 0.44 \\ 0.18 & 0.15 & 0.12 & 0.11 & 0.44 \\ 0.18 & 0.15 & 0.12 & 0.11 & 0.44 \\ 0.18 & 0.15 & 0.12 & 0.11 & 0.44 \end{pmatrix},$$

This matrix **F** indicates that the collective opinion was formed due to the contribution of each member of the group, and the leader's contribution prevails but is not decisive. As noted in the paper (Homans 1958) in a theoretical analysis of group behavior, advancement to a leadership position is ultimately determined by the effectiveness of the contribution of a group member to the solution of a group problem. It is clear that in this case, the consensus becomes more harmonious and balanced.

Above, the conditions and stages of modeling were given, which we will adhere to when modeling the case of the presence of a leader in the group. We will assume that he gives equal trust to all members of the group, i.e., the simulation matrix **P** has the form:

$$\mathbf{F} = \begin{pmatrix} p_{11} & 0 & 0 & 0 & 1-p_{11} \\ 0 & p_{22} & 0 & 0 & 1-p_{22} \\ \dots & \dots & \dots & \dots & \dots \\ 0 & 0 & 0 & p_{n-1,n-1} & 1-p_{n-1,n-1} \\ 1/n & 1/n & 1/n & 1/n & 1/n \end{pmatrix}$$

In this case, the number of iterations (negotiations) is on average greater than in the case of the presence of a dominant member in the group. Just as in the case of a leader in a group, there is a hyperbolic relationship between the number of iterations (negotiations) and the average authoritarianism of group members. The regression dependence has the form $\hat{m} = \frac{9.2}{1-p}$, where is the average number of iterations (negotiations), p is the average authoritarianism of group members; the coefficient of determination

$R^2 \approx 0{,}96$ reflects the 96% contribution of the influence of the group members' authoritarianism to the model.

The simulation results showed a more pronounced variability in the number of approvals in the presence of a leader compared to a domination member, all other factors being equal: in the conditions of an authoritarian member, the standard deviation was about $\sigma \approx 1.7 \div 1.8$ approvals per 100, and in the conditions of leadership, the standard deviation reached $\sigma \approx 23.8$ approvals per 100. Thus, on average, in the presence of a leader, the time to reach consensus is delayed, although not significantly, compared to an authoritarian member and is subject to greater variability.

It is known in social psychology that leadership as a phenomenon is less stable than leadership and depends on group approval: the high authoritarianism of group members leads to a decrease in leadership positions in relation to the team, thereby setting a less predictable process in terms of the number of approvals before the formation of a collective opinion.

But, as noted above, if there is a leader in the group, a collective decision is formed considering the contribution of the opinion of each member of the group. Thus, it is confirmed that leadership is a more complex system of relations in a group, in contrast to domination, which was noted earlier in the field for small groups.

7.7 Influence of the Unilateral Concession Upon Variability of the Number of Negotiations

At simulation of a consensus under the condition of coalitions, the issue of the influence exerted by various factors upon the average number of negotiations performed until the consensus was built, was considered in the previous sections. It is noted that different a variability in the number of negotiations is observed under different conditions. Variability provides for an important understanding of the boundaries of variation in the number of negotiations in the coalitions because there is no complete homogeneity inside a coalition. Therefore, it characterizes the constancy of the number of steps taken to build a consensus at fixed levels of the predetermined factors. If the variability is expressed strongly, it provides us with the reason to seek new factors explaining the variability that would, in its turn, allow forming of the conditions for intragroup stabilization of the group members and further investigation of the model under the given conditions.

Figure 6 provides the box-and-whiskers diagrams of the variability of the number of negotiations at the pre-defined parameters of the concession and the strength index of coalitions. Increasing both interquartile scattering and the extremum is notable at the transition from a large concession

$Y = 75\%$ to a small one $Y = 75\%$ (Fig. 10). A larger concession can be interpreted as a method for stabilization of the number of negotiations at different modes of simulation. In the case with a large concession, all other possible factors characterizing the group do not result, at the pre-defined index I, in any essential variation of the time elapsing until the consensus is built.

At a small concession, this stabilization is breached under all other equal conditions. Practically, it means that confidence in the number of negotiations m predicted according to the formula (6) is lost for a specific group. Variation of the mean squared deviation (MSD) of the number m occurs in a nonlinear manner from a small concession $Y = 10\%$ to a large one $Y = 75\%$ irrespective of the dependence on the ratio of the number of members between the coalitions I (Fig. 10). At a concession of $Y \geq 60\%$ the value of MSD is ≤ 1, i.e., there occurs a stabilization of the number of negotiations from one simulation to the other.

During the simulation, the largest scattering among the number of negotiations in the experiments (Fig. 11) is observed at the concession of $Y = 10\%$. It can be assumed that such variability is stipulated by the interaction of the members both within each of the coalitions and between separate coalitions. The correlation analysis revealed a significant relationship ($r > 0.5$) of the number of negotiations with the maximum authoritarianism and the range of the confidence levels of the coalition accepting the concession (among all of the considered factors reflecting the structure of coalitions).

To explain that variability in detail, the cluster analysis is performed in the *Statistica* 15 software module with the following assumptions (Fig. 12):

(1) considering that the power index of coalitions I exerts no influence upon the value of MSD, the value taken for the cluster analysis $I = 1$ is that corresponds to the equal size coalitions; the concession is $Y = 10\%$;

(2) the clustering was performed based on three factors: the number of negotiations m, the highest authoritarianism among the members of the coalition accepting the concession, and the range R calculated among all the probabilities of the confidence of members of the coalition accepting the concession.

(3) the Euclidean distance was selected as the clustering metric.

The highest values of authoritarianism and range among the probabilities of confidence of the members of the coalition characterize the degree of negotiations of its members during the decision-making process. Considering that the consensus-based decision will be ultimately determined by just the opinion of the coalition, which accepts the

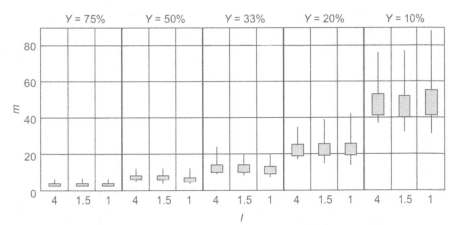

Figure 10: Box-and-whiskers diagrams for the number of negotiations m at the pre-defined power index of coalitions I and different concessions Y.

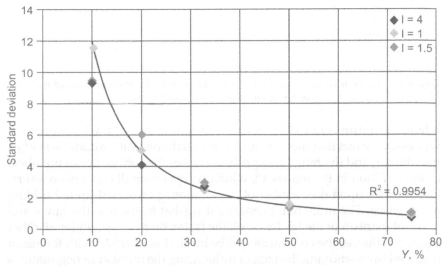

Figure 11: Dependence of MSD of the number of negotiations upon the concession Y at different power indices of coalitions I.

concession, then it is quite clear that it is its structure, which is of primary importance. The performed cluster analysis separated three large groups, in which the data are collected—with a small number of negotiations, a large and a medium number thereof (please see the values on the vertical axis in Fig. 12).

Detail analysis of the structure of each cluster revealed the following interrelation: there exists a notable and significant correlational relationship ($r \approx 0.5 \div 0.6$) within the clusters with a small (not more than 43) and a large (over 60) number of negotiations; for a medium number group that

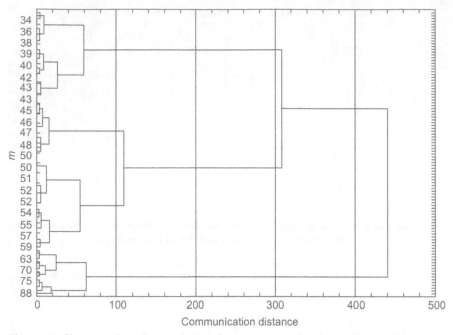

Figure 12: Cluster analysis for simulation of a consensus under the conditions of the equal size groups for a group of 20 members.

relationship turns out to be weak and insignificant—at the level of 5%. It witnesses the fact that the structure of the coalition (authoritarianism of its members p_{ii} and the range R of probabilities of the confidence of members of the coalition in the matrix **P**), which accepts a small concession, exerts an influence upon the number of negotiations performed up to building a consensus. The more homogeneous it is (that is, the less the range and authoritarianism of the leaders are), the faster, on average, under all other equal conditions, the consensus will be built. It is matched with the result obtained from studying the factors influencing the number of negotiations for the technical committee without forming coalitions (Aronov et al. 2018).

7.8 Main Results

Result 1: An important distinctive feature of the first model is the fact that it considers the consensus-building time dependency on the number of the group and the authoritarianism of its members. Also, in the first approach control is introduced, due to which it is possible to correct (to reduce) the number of agreement stages needed for consensus-building.

The increase in the number of the group, the same as the increase of authoritarianism of the group members, negatively affects the consensus-building time and, hence, the work effectiveness of the group.

For achieving consensus-building a planned time control can be introduced. This control is to exclude situations (global dominance, presence of several leaders, responsibility shift, coalitions), in which consensus is fundamentally unattainable it will take significant time to attain a consensus.

It is worth noting that the result obtained can be extended to any organisational structure, in which decision is made based on consensus: the increase of the number of the group in these structures and the increase of authoritarianism of its members significantly complicate the consensus achievement.

Result 2: "Weeding" of the **P** matrix main diagonal can be interpreted as a transition to decision-making by the method of consensus minus one, consensus minus two, etc., which in practice is used in the development of "incomplete consensus" documents.

Result 3: The opinion structure affects the consensus-building time. The agreement process, on averagely, is delayed in the case of one authoritarian member with an uncontested opinion rather than in the case of one authoritarian member with a detailed opinion. In practice, it means that it is much more difficult to find common points of contact with such a person in a question posed. This is consistent with a well-known fact in sociology (see, for example, Moscovici 1976).

Result 4: Leadership is a more complex system of relations in a group, in contrast to domination, which was noted earlier in the field for small groups. The analysis showed that the presence of a dominant member in the group significantly reduces the variability in the number of iterations to consensus. In this case, the variability in the number of approvals is minimal, but the consensus group decision, as noted earlier, is provided only by the opinion of the dominant member. Therefore, from a practical point of view, the appearance/ formation of a leader in a group provides a high-quality and relatively fast consensus decision (in comparison with a homogeneous team).

Result 5: Consensus can always be built at a unilateral concession in the model with coalitions. Whereas, the less value the concession is, the more negotiations will be required to build a consensus. Even a small concession, i.e., striving to find a compromise, results in the occurrence of a consensus.

Increasing the compromise size results (under all other equal conditions) in a sharp decrease of the time elapsing until the consensus is built.

It is shown that the time elapsing until the consensus is built is poorly dependent on the "strength" of the coalition. Whereas a concession of a larger coalition with less numerous coalitions does not require more time to elapse until the consensus is built than otherwise. Practically it is stipulated primarily by a different mode of interaction between the coalitions in the group at different types of concession: at a concession from a small coalition, a larger coalition "pulls over" the less numerous one due to a larger number of the members, while to force a large coalition to make a concession it would be necessary that a small coalition would provide a strong argumentation of its position. It is demonstrated that the eigen demands of the party are decreased while conceding to provide the other side with the opportunity to attain the desired result.

In the case when the coalition consists of one authoritarian leader, to increase the quality of consensus and decrease the time elapsing until the consensus is built, it would be reasonable for the authoritarian leader to make a concession. In this case, the consensus-based decision turns out to be more balanced.

Studied variability of several negotiations until building a consensus showed that the structure of the coalition accepting a small concession exerts an influence upon several negotiations until building a consensus. The more homogeneous the coalition is, the faster, on average, the consensus will be built under all other equal conditions. At a larger concession stabilization of the number of negotiations occurs, and the initial structure accepting the concession loses its significance for the number of negotiations.

Based on the known references social psychologists failed to study the time spent in the search for consensus under the conditions of coalitions; therefore, the obtained results might be rather interesting for this sphere of research. The obtained results can be distributed in any organizational unit, in which the decisions are made based on building a consensus: the creation of coalitions in such organizations results in the impossibility of the decision-making process within the framework of building a consensus and requires consideration of various methods used for taking a consensus-based decision.

7.9 Conclusion

The main problems of reaching consensus in a social group are analyzed using a mathematical model of consensus based on regular Markov chains, and consensus modeling. It is shown that an increase in the number of members of a group and their authoritarianism negatively affect the time to reach a consensus. The proposed mathematical model of consensus is adapted for the case of two coalitions in a group when consensus is achieved through a concession. It is proved that the time to reach of consensus significantly depends on the size of the unilateral concession and weakly depends on the number of coalition members. It was revealed that even a small concession leads to a consensus. Consensus modeling shows: that the concession of the authoritarian leader in the group reduces the time of the negotiation process and improves the quality of the consensus; deciding according to the «consensus minus one» rule significantly reduces the time to reach consensus (up to 97%), which makes it possible to recommend this rule for practical application.

References

Aronov, I. Z., O. V. Maksimova, and V. I. Grigoryev. 2018. Analysis of consensus-building time in social groups based on the results of statistical modeling. *In Advanced Mathematical Techniques in Science and Engineering*, 1–31. Netherlands: River Publishers.

Aronov, I. Z., and O. V. Maksimova. 2020. Theoretical modeling consensus building in the work of standardization technical committees in coalitions based on regular Markov chains. *Computer Research and Modeling* 12(5): 1247–56.

Aronov, I. Z., A. M. Rybakova, and N. M. Galkina. 2021. Peculiarities of technical measures during the COVID-19 pandemic. *In Use of AI, Robotics, and Modern Tools to Fight Covid-19*, 141–159. Netherlands: River Publishers.

Chebotarev, P. 2010. Comments on "Consensus and cooperation in networked multi-agent systems." *Proc. IEEE* 98(7): 1353–1354.

Della Porta, D., M. Andretta, L. Mosca, and H. Reiter. 2006. *Globalization from below: transnational activities and protest networks*. Minneapolis: University of Minnesota Press.

DeGroot, M. H. 1974. Reaching a consensus. *J. Am. Stat. Assoc.* 69: 118–121.

Efron, B. and R. Tibshirani. 1991. Statistical data analysis in the computer age. *Sci. New Ser.* 253(5018): 390–395.

Gantmacher, F. R. 1959. *The Theory of Matrix*. New York, NY: Chelsea Publishing Company, 1959.

Gelderloos, P. 2006. *Consensus: A New Handbook for Grassroots Political, Social and Environmental Groups*. Tucson, AZ: See Sharp Pr.

Gençer H. 2019. Group Dynamics and Behaviour *Universal Journal of Educational Research* 7(1): 223–229.

Gockel, C., N. L. Kerr, D. H. Seok, and D. W. Harris. 2008. Indispensability and group identification as sources of task motivation. *Journal of Experimental Social Psychology*, 44(5): 1316–1321.

Homans, G. C. 1958. Social Behavior as Exchange. *American Journal of Sociology* 63(6): 597–606.

Kerr, N. 1989. Illusions of efficacy: The effects of group size on perceived efficacy in social dilemmas. *Journal of Experimental Social Psychology* 25: 287–313.

Kemeny J. G., and J. L. Snell. 1960. *Finite Markov chains*. Princeton: The University Series in Undergraduate Mathematics.

Kozyakin, V., N. Kuznetsov, and P. Chebotarev. 2019. Consensus in Asynchronous Multiagent Systems. I. Asynchronous Consensus Models. *Automation and Remote Control* 80(4): 593–623.

Kudish, S., S. Cohen-Chen, and E. Halperin. 2015. Increasing support for concession-making in intractable conflicts: The role of conflict uniqueness. Peace and Conflict. *Journal of Peace Psychology* 21(2): 248–263.

Mazalov, V., and Y. Tokareva. 2012. Arbitration procedures with multiple arbitrators. *Eur. J. Oper. Res.* 217: 198–203.

Moscovici, S. 1976. *Social Influence and Social Change*. New York, NY: Academic Press.

Myers, D. 2015. *Social Psychology*. New York: The McGrow-Hill Companies, Inc., 2010.

Open science collaboration. Estimating the reproducibility of psychological science. *Science* 349(6251).

Seungwoo, K., and L. R. Weingart. 2004. Unilateral Concessions from the Other Party: Concession Behavior, Attributions, and Negotiation Judgments. *Journal of Applied Psychology* 89(2): 263–278.

Sheril, L., L. Johncon, J. Leedom, and L. Muhtadie. 2012. The dominance behavioral system and psychopathology: evidence from self-report, observational, and biological studies. *Psychol. Bull* 138(4): 692–743.

Tukey, J. W. 1977. *Exploratory data analysis. Reading.* MA: Addison-Wesley.

Zazhigalkin, A. V., I. Z. Aronov, O. V. Maksimova, and L. Papic. 2019. Control of consensus convergence in technical committees of standardization based on regular Markov chains model. *International Journal of Systems Assurance Engineering and Management* 1: 1–8.

CHAPTER 8

Data to Data Science
A Phenomenal Journey

Mohammad Haider Syed,[1,*] *Sidhu*[2] and *Kamal Upreti*[3]

Contents

8.1 Introduction

Since the time of human evolution, data is of utmost importance from counting to barter and other related activities. These data with human wisdom have meaningful information. Data in early history was inscribed on stone and then with the invention of paper on paper. Furthermore, with the invention and advancement of electronic gadgets and digitization, larger quantities of data started getting generated and

[1] Department of Computer Science, College of Computing and Informatics, Saudi Electronic University, KSA.
[2] Department of Mathematics, Shri Venkateshwara, University, Uttar Pradesh, India.
[3] Department of Information Technology, Indraprastha Engineering College, Ghaziabad, Uttar Pradesh, India.
* Corresponding author: m.haider@seu.edu.sa

these generated data were stored. These digitally stored data are then churned to extract meaningful information (Tukey 1962, 1984). In 1962, J.W. Tukey predicted that the advancements in computing power will be facilitating data analysis as an empirical science. Later in 1977 (Tukey 1977), the author emphasized the use of data for the hypothesis test and data analysis (Naur 1974). The book explored the various methods and techniques of data processing that are used in different applications. Formally "Data is defined as the representation of fact and/or idea in a formal manner to be communicated and/or manipulated by some process to extract information". The term "Data Science", has been extensively used in Naur (1968). The definition of data science coined by Naur is, "The science of dealing with data, once they have been established, while the relation of the data to what they represent is delegated to other fields and sciences". In 1989 the first Knowledge Discovery in Databases workshop was organized. Later in the year 1995 onwards, it became a regular annual affair by the name of ACM SIGKDD. The advancement of technology over time has given the flexibility to humans to generate and capture data at a very high speed. These generated data are captured efficiently and stored to be churned to know something non-trivial. May this be capturing the buying behavior of the customer or surfing the web pages over the internet? So, it is worth saying that data is available all over, but we are starving for the information out of the generated or stored data. So, in 1996 the term data science was included in the conference title "Data science, classification, and related methods". In (Fayyad et al. 1996) proposed that knowledge discovery from Databases can be referred to by many names like data dredging, information processing, etc. Thus, it can be referred to as discovering useful information from the data and the process of extracting useful patterns or knowledge is termed data mining. The process of extracting meaningful information from the stored data by applying a mining algorithm was limited to not a very large data set and it worked well for small data sets. But with the massive rate of data growth with time and its storage needs something new to churn and extract useful information from this mammoth amount of stored data. Breiman (2001) proposed that data can be generated by two approaches firstly by stochastics processes and secondly by some heuristic approach, but for data generated by a heuristic approach its mechanism is not known. The approach adopted by statisticians remains confined to the known current problems whereas, the heuristic approach of data has gained popularity in the much wider field of application and problem-solving. With the evolution of modern industrialization, the rate of data generation has increased many folds and thus there is a huge amount of data available. This stored data which is mammoth in size has many issues associated with it. Section 2 discusses warehousing followed by its challenges in

Section 3. Section 4 talks about data mining and its related issues. Section 5 is about big data its issues and challenges discussed in Section 6, followed by a conclusion in Section 7.

8.2 Data Warehousing

This available stockpile of data needs to be pre-processed. The technique of collecting the data from heterogeneous, autonomous and possibly located different geographical locations and its processing has been referred to as Data Warehousing (Inmon 1995). Formally data warehouse is defined as a "Collection of data, which is subject-oriented, integrated, time-varying, and non-volatile to support decision making".

Voluminous data collected and stored has some associated challenges that need to be addressed. Such as this data is dirty, and this dirty data needs to be cleaned to have unbiased results. Cleaning dirty data itself is a challenging task. This involves filling missing values, incomplete data, data collected using faulty instruments, human error, default data, etc. These activities still are a major challenge to the field of data science.

Figure 1 shows the abstract view of the data warehouse as to how data at the base sources are monitored. If some changes are noticed the newly added data is then extracted from these base sources and wrapped into a common format and integrated. This integrated data is then stored in the archival repository referred to as a data warehouse. But the process of collecting, integrating and storing is a challenge.

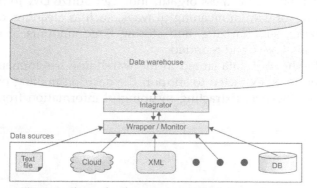

Figure 1: Shows the abstract view of the data warehouse.

8.3 Data Warehousing and its Issues

Using technologies like Data Warehousing and Business Intelligence (BI) analytical tools, companies can improve their performance and competitive positions by boosting their capacity to respond swiftly to rapid environmental changes with high-quality business decisions. Nonetheless, having relevant, trustworthy, accessible, accurate, timely, comprehensive, coherent, and consistent quality information for the decision at hand is essential for ensuring business intelligence. As a result, improving business intelligence through better decision-making is a crucial priority for today's corporate leaders.

Without a question, accessing a significant and massive amount of data kept in a business's operational systems has become increasingly time-consuming and inconvenient. As a result, businesses have turned to data warehousing to solve these issues by merging disparate operational data sources. The research community is still at large working to get something concrete on these challenging issues (García et al. 2015, García et al. 2016, Ramírez-Gallego et al. 2017, Alexandropoulos et al. 2019). As shown in Fig. 2 still researchers and data scientists are struggling with these issues. Google search trend is shown in Fig. 2 for searches like data cleaning, data integration, data transformation and data reduction. The data integration (Lenzerini 2002, Halevy et al. 2006, Doan et al. 2012) is one of the most challenging issues to date followed by data cleaning and data transformation. The google search trend shown in Fig. 2 is for the last five years.

Large amounts of data are cleaned, aggregated, and condensed in a multidimensional data structure to facilitate multidimensional analysis in DWs. Not only does a DW identify the need for current and future data, but it also recognizes the necessity for previous data. For example, trend analysis, regardless of company size, necessitates a large amount of historical data. A business organization can utilize DW to manipulate vast amounts of data in meaningful ways, such as cleaning, organizing, characterizing, summarizing, and storing large amounts of data to be converted, analyzed, and reported.

This Plethora of data stored in the warehouse has immense hidden information to be extracted to support the decision-making process and business intelligence. Extracting meaningful information from the data

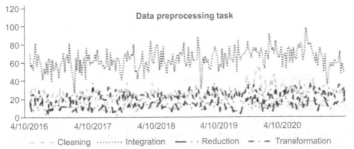

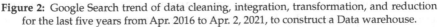

Figure 2: Google Search trend of data cleaning, integration, transformation, and reduction for the last five years from Apr. 2016 to Apr. 2, 2021, to construct a Data warehouse.

warehouse which is voluminous in size also has big challenges. All the data collected from the base sources are stored in the warehouse and may not be needed to facilitate the process of decision-making. The query posed on the data warehouse is complex and exploratory as it involves lots of aggregates and joins. In response to the query posed only a small portion of the data from the entire data warehouse is used. Even for only a small portion of data that is needed to construct the query response entire data warehouse will be explored. As queries posed on warehouses are complex and exploratory so the query response time is very high (Gupta 1997, Gupta and Mumick 2005, Scheuermann et al. 1996). This challenge was addressed by materializing the views (Gupta and Mumick 1995, Chaudhuri et al. 1995, Kumar and Haider 2010, Haider and Kumar 2017, Haider and Vijay Kumar 2011). Viewing materialization from the existing data warehouse involves view evolution, view materialization, and view maintenance. This problem has been classified as the NP-class of problem (Harinarayan et al. 1996).

The collection of data from the base sources does not guarantee that it will be useful for a decision support system. The primary focus of the data collection and preparation should be to populate data to solve the problems. Thus problem-oriented data collection improves the accuracies of the data mining models many folds (Pyle 1999, Zhang et al. 2003). Data collected and stored in mammoth warehouses may not serve the purpose as it may not be collected focusing on the specific problem. Assembling the data in the data warehouse in the meanable form is also a matter of concern. So that this data can be used for mining useful hidden, non-trivial pieces of information. This approach of extracting hidden, implicit, and non-trivial information for the warehouse is referred to as data mining.

8.4 Data Mining and its Challenges

In data mining, a heuristic approach is applied to a huge amount of archival data to extract meaningful patterns. These patterns are then studied, analysed, interpreted, and concluded with human wisdom. A lot of research in the area of data mining has been done and is continually explored by researchers to improve. Figure 3 shows a research taxonomy of the pattern mining (Han et al. 2011). Despite the importance of data warehousing technologies in theory and practice, few studies of data warehousing problems and success factors have been done. Although it appears that many businesses have embraced DW, the road to success has been littered with setbacks. Many additional societal/cultural, organizational, and technical factors, in addition to the high cost and resource needs for such projects, may be to blame for their high failure rates.

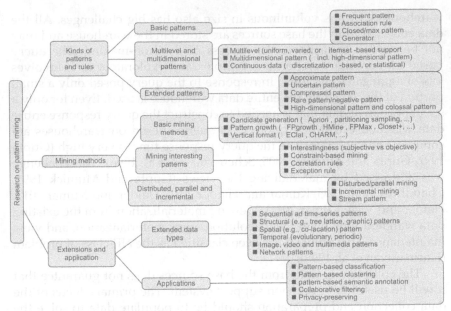

Figure 3: Research taxonomy of pattern mining.

Before applying any mining technique, algorithm data is prepossessed and framed as per the requirements of the data mining query. Major goals of the data mining tasks can be summarized as data pre-processing, association, classification, prediction, clustering, etc.

As discussed earlier, one of the challenges of the data mining task is to clean the data that needs to be processed for a specific purpose. Pre-processing is done for the following reasons missing value, variance in data, noise, etc. Once the results in response to the data mining query are returned by the OLAP engine, interpretation of this pattern is also one of the major challenges in data mining. A number of the mining algorithms available are good in numbers, so these algorithms also add complications as to which algorithm to use for specific situations and what will be the result produced by the algorithm. Data may be geographically located at different places and may/may not be autonomous, but heterogeneity adds another level of challenge. The data collected from these sources need to be processed for data quality. Interpretation of the mined result is also one of the major issues with the data miners as most of the available algorithms were not able to differentiate between co-occurrence and casualty (Yang and Wu 2006, Reddy 2011).

Often when predicting the future, the result set produced by the mining algorithm is large thus it becomes difficult to interpret the pattern. Other major issues that need to be addressed are scalability, dimensionality, data

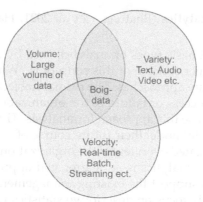

Figure 4: 3 V's of Big data.

stream, distributed data, etc. As the data mining process uses archival data, the rate at which stream, volume, and variety of data are generated needs an alternative approach. This was referred to as Big-Data, a term coined in 2005 by Roger Mougalas. Big data refers to 3-V namely volume, velocity, and variety of data over the internet. Thus, big data is an amalgamation of these 3 v's as shown in Fig. 4.

8.5 Big Data

Advances in technology and deep penetration of this technology in human life across the globe and space have led to the exponential growth of data. Data which are generated in large volumes at a very high velocity and variety has made data mining a challenging task. These quintillion bytes of data generated in a non-structured form like text, images, audio, social posts, sensors, etc., are to be captured and processed in real-time. This tsunami of data from various sources also needs critical analysis of what to keep and what to discard. This science has influenced almost all walks of life from retail marketing to space science. This approach has a wide variety of applications may it be improving the quality of education, environment modelling, or minimizing the cost of healthcare with the improved quality. This has motivated us to analyse different kinds, and types of data. Meticulous use of data by organizations to gain the right information which knows will empower the companies to grow and have a competitive edge over the others. Big data analytics cannot be used as a single window strategy for all kinds of decisions. Literatures (Greasley 2019, Deka 2014, Menezes et al. 2019) proposed that there are different types of analytics for different purposes. These analytics are namely Descriptive, Predictive and Perspective. These offer analytics of different kinds to organizations. Each of these is discussed below.

a. **Descriptive Analytics** (Bhatnagar et al. 2021, Han et al. 2020, Nida 1949)

It is one of the most basic types of analytics used by the organization for basic analysis. A major concern of this concept is "What has happened?", this will lead to future approaches. The primary objective of this is to find the reason for the success or failure of the organization in the past. Thus, most organizations use this approach for analysis. This historical analysis helps organizations to model their future course of action. Thus, it can be used to get an aggregated overview of the organization's performance over time. As descriptive analytics is an aggregated approach, it summarises the existing data to support the existing management information tools. These tools primarily focus on descriptive statistics which involves basic arithmetic operations. As an example, many E-learning systems use this technique to analyze the assessment of assignments and grades, comparing the results, time taken to complete the course, etc.

b. **Predictive Analytics**

This technique analyses the past trend of data and predicts future trends for the organizations. Thus, the major focus is on "What could happen in future". They enable the organizations to realize the realistic goal, effective planning and in controlling exaggerated expectations. This futuristic approach to analytics strengthens the likelihood of future events using machine learning algorithms. The approach has been further categorized as Predictive modelling, Root Cause Analysis, Forecasting, Pattern identification etc. Sentiment analysis uses the concept of predictive analysis (Junqué de Fortuny et al. 2013, Ratner 2017, Kumar et al. 2021). Some examples of predictive analysis are forecasting, customer behavior, etc.

c. **Perspective Analytics** (Kumar 2021, Deshpande et al. 2019)

This analytics is a compliment to predictive analysis that acts as a catalyst to manipulate the future. In other words, perspective analysis proposes the action that may facilitate improving the business metrics. In a nutshell, this approach can be used to simulate and optimize the business process. Thus, it can be referred to as analytics of optimization to accelerate the future growth of the organization. To simulate the future of the organization, the Perspective analysis uses the permutation and combination of data and various business rules. This concept of analytics is not very popular among organizations as its form is complex. It uses the concept of Machine learning, NLP, operation research, etc.

d. **Diagnostic Analytics** (Belle et al. 2015, Deshpande et al. 2019)

To understand the "Why" behind what has happened the concept of Diagnostic analytics is used. This gives an in-depth insight into the problem set. Thus, anomalies and their causal relationship in the data can be determined.

These analytical approaches are useful if the data used for the analysis is of good quality. Getting good quality data has many challenges, as data is semi-structured or unstructured.

8.6 Challenges of Big Data

Data from the different sources are not in a structured form rather they are in a mixed form of structured, semi-structured and unstructured forms. So collectively these data can be referred to as semi-structured as audio and video are structured but other data is non-structured. Thus retrieval, organization, modelling, and analysis is a major challenges. Though many benefits are counted on big data at the same time, major challenges (Sivarajah et al. 2017, L'heureux et al. 2017, Chen et al. 2016, Barnaghi et al. 2013) are also to be addressed with equal importance. As shown in Fig. 4, not only the volume of data is a challenge but also the velocity and variety have equal challenges to be addressed. Volume, velocity, and variety of data generated need to be collected. Collecting this diversified and semi-structured data is a major challenge for the researchers. As all the data may not be useful for analysis and filtering and pre-processing are still major issues with the research community. Extraction and cleaning the useful data are part of pre-processing and it requires a mature approach.

Data which is generated from heterogenous sources and architecture needs to be mapped to a common format before processing. This also requires aggregation and proper representation. These tasks need to be automated and a scalable approach needs to be adapted to have a good quality of data. Once data is assimilated a suitable query is also needed as data collected can be noisy, heterogenous and untrustworthy. So, a robust technique needs to be drawn to make this data clean, noise-free and trustworthy. Analyzed data needs proper interpretation and for this, all kinds of assumptions need to be considered as the results may be an error because of many reasons like human error, instrument error and so on. These advancements have opened a plethora of opportunities but at the same time, a lot more needs to be done to harness its real potential.

Big data challenges can be divided into three categories. Data challenges are the first type, followed by data process challenges, and finally data management. The issues related to the properties of big data are known as data challenges. Process challenges are those that arise during the data processing process, whereas management challenges are those that arise when dealing with data, such as ensuring security. Big data's properties, such as its large amount and variety, provide numerous issues. Data collecting, pre-processing, data analysis, and data visualization are all process difficulties, whereas privacy and security are management challenges.

Big data challenges can be divided into three categories. Data challenges are the first type, followed by data process challenges, and finally data management. The issues related to the properties of big data are known as data challenges. Process challenges are those that arise during the data processing process, whereas management challenges are those that arise when dealing with data, such as ensuring security. Big data's properties, such as its large amount and variety, provide numerous issues. Data collecting, pre-processing, data analysis, and data visualization are all process difficulties, whereas privacy and security are management challenges.

The following are some of the most prominent big data challenges:

The first difficulty is volume; the tremendous increase in data from both internal and external sources has resulted in a massive amount of data. This large volume of data poses issues to the data itself, such as the inability to store data for processing using traditional techniques, necessitating the development of more imaginative approaches to deal with the data flood. Variety's variety challenge is tied to the various shapes it takes. Structured, semi-structured, and unstructured data can all be found in huge amounts. According to research, 95 per cent of data is stored in an unstructured format. As a result, putting it into a form that can be used for analysis is a significant difficulty. The velocity challenge refers to the rate at which data is generated by the devices. Batch processing and real-time processing are two methods for processing data. Data is stored and then processed in batch processing, whereas real-time processing is ongoing. Real-time processing is necessary for online shopping to provide value to customers. The task is a data veracity challenge, in which data veracity refers to the data's quality and accuracy. It is concerned with data fabrications, imprecision, chaos, and misplaced evidence. When a crucial judgment must be made, it defines the data's credibility. User opinions on social networking sites can be characterized as good, negative, or neutral. One of the most important components of big data characteristics is value challenges. Big data is full of useful information that must be pulled from massive datasets. This poses a significant challenge to data, such as extracting high-value information from data in a cost-effective manner and applying it to business intelligence, health care, and other areas.

Processing and interpreting huge datasets are procedural difficulties. It adds a huge obstacle to the process because the data is in various formats and converting it into a single format for analysis is a difficult effort. It is broken into four sections: Data Pre-processing, Data Analysis &

Modeling, and Data Visualization are all examples of data acquisition and storage. Obtaining and storing information Data acquisition is the process of gathering and storing information for future use. Data is collected from a variety of sources, including sensors, social networking sites, blogs, and other sources, and as a result, data is available in a variety of formats (structured, semi-structured, and unstructured), posing a substantial issue.

The second issue is storage, as data created by various devices does not necessarily mean that the entire data is meaningful, so a smart filter must be used to generate useful datasets. To handle this enormous dataset, high-cost scalable systems may be required. Data pre-processing is the procedure for extracting high-quality data from big datasets, as poor data leads to poor knowledge. As a result, data preparation is crucial in the discovery of new knowledge. Before applying big data mining techniques to the data, noise, missing values, inconsistent and unnecessary data, and so on are removed. The Feature Selection approach receives the majority of the attention in big data pre-processing, whereas other methods, such as reduction, missing value imputations, and noise treatments, are often overlooked.

Figure 5 shows the growing demand for data science as per the google search trend. This data is captured for the period of the last sixteen years and it can be seen from the figure, that there has been phenomenal growth in the area.

In Fig. 6 the Google search trends of the term data mining, big data, and data science are shown. The search for data science has outnumbered the other two approaches namely data mining and big data.

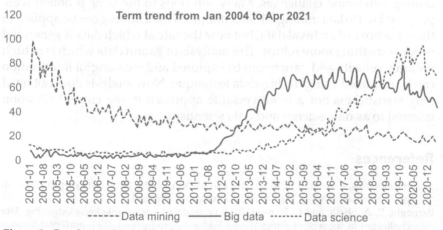

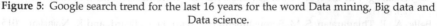

Figure 5: Google search trend for the last 16 years for the word Data mining, Big data and Data science.

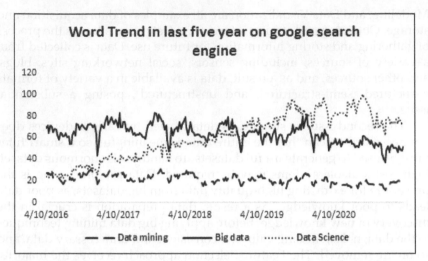

Figure 6: Google trend of data mining, big data, and data science for the last five years.

8.7 Conclusion

This chapter provides a brief overview of data warehousing. How it evolved and its associated challenges, and the necessary steps taken to prepare this archival repository of data. This repository of archival data can be used to support the decision-making process. The analysis of data includes data quality, data cleansing, data integration, data selection, data transformation, pattern evaluation, knowledge presentation and data mining issues and challenges. Many solutions to the large problem were proposed with data mining. This technique of data mining can be applied to the repository of archival data, but now the rate at which data is generated needs something more robust. This analysis of gamut data which is of high volume, velocity and variety can be explored and meaningful information can be extracted using the big-data technique. Now analysis does not need only statisticians but a more versatile approach in the current situation referred to as data science and data scientist.

References

Alexandropoulos, S.-A. N., S. B. Kotsiantis, and M. N. Vrahatis. 2019. Data preprocessing in predictive data mining. *The Knowledge Engineering Review* 34.

Barnaghi, P., A. Sheth, and C. Henson. 2013. From data to actionable knowledge: Big data challenges in the web of things [Guest Editors' Introduction]. *IEEE Intelligent Systems* 28: 6–11.

Belle, A., R. Thiagarajan, S. M. Soroushmehr, F. Navidi, D. A. Beard, and K. Najarian. 2015. Big data analytics in healthcare. *BioMed Research International*.

Bhatnagar, V., R. C. Poonia, P. Nagar, S. Kumar, V. Singh, L. Raja, and P. Dass. 2021. Descriptive analysis of COVID-19 patients in the context of India. *Journal of Interdisciplinary Mathematics* 24: 489–504.

Breiman, L. 2001. Statistical modeling: The two cultures (with comments and a rejoinder by the author). *Statistical Science* 16: 199–231.

Chaudhuri, S., R. Krishnamurthy, S. Potamianos, and K. Shim. 1995. Optimizing queries with materialized views. *In Proceedings of the Eleventh International Conference on Data Engineering*, 190–200. IEEE.

Chen, Y., J. D. E. Argentinis, and G. Weber. 2016. IBM Watson: how cognitive computing can be applied to big data challenges in life sciences research. *Clinical Therapeutics* 38: 688–701.

Deka, G. C. 2014. Big data predictive and prescriptive analytics. *In Handbook of Research on Cloud Infrastructures for Big Data Analytics* (IGI Global).

Deshpande, P. S., S. C. Sharma, and S. K. Peddoju. 2019. Predictive and prescriptive analytics in Big-data Era. In, *Security and Data Storage Aspect in Cloud Computing* (Springer).

Doan, A. H., A. Halevy, and Z. Ives. 2012. *Principles of Data Integration* (Elsevier).

Fayyad, U., G. Piatetsky-Shapiro, and P. Smyth. 1996. From data mining to knowledge discovery in databases. *AI Magazine*,17: 37.

García, S., J. Luengo, and F. Herrera. 2015. *Data Preprocessing in Data Mining* (Springer).

García, S., S. Ramírez-Gallego, J. Luengo, J. M. Benítez, and F. Herrera. 2016. Big data preprocessing: methods and prospects. *Big Data Analytics* 1: 1–22.

Greasley, A. 2019. *Simulating Business Processes Ffor Descriptive, Predictive, Aand Prescriptive Analytics* (De Gruyter).

Gupta, A., and I. S. Mumick. 1995. Maintenance of materialized views: Problems, techniques, and applications. *IEEE Data Eng. Bull.* 18: 3–18.

Gupta, H. 1997. Selection of views to materialize in a data warehouse. *In International Conference on Database Theory*, 98–112. Springer.

Gupta, H., and I. S. Mumick. 2005. Selection of views to materialize in a data warehouse. *IEEE Transactions on Knowledge and Data Engineering* 17: 24–43.

Haider, M., and T. V. V. Kumar. 2011. Materialised views selection using size and query frequency. *International Journal of Value Chain Management* 5: 95–105.

Haider, M., and T. V. V. Kumar. 2017. Query frequency based view selection. *International Journal of Business Analytics (IJBAN)* 4: 36–55.

Halevy, A., A. Rajaraman, and J. Ordille. 2006. Data integration: The teenage years. In *Proceedings of the 32nd International Conference on Very Large Data Bases*, 9–16.

Han, J., M. Kamber, and J. Pei. 2011. Data mining concepts and techniques third edition. *The Morgan Kaufmann Series in Data Management Systems* 5: 83–124.

Han, Y. N., Z.-W. Feng, L.-N. Sun, X.-X. Ren, H. Wang, Y.-M. Xue, Y. Wang, and Y. Fang. 2020. A comparative-descriptive analysis of clinical characteristics in 2019-coronavirus-infected children and adults. *Journal of Medical Virology* 92: 1596–602.

Harinarayan, V., A. Rajaraman, and J. D. Ullman. 1996. Implementing data cubes efficiently. *Acm Sigmod Record* 25: 205–16.

Inmon, W. H. 1995. What is a data warehouse? *Prism Tech Topic* 1: 1–5.

Junqué de Fortuny, E., D. Martens, and F. Provost. 2013. Predictive modeling with big data: is bigger really better? *Big Data* 1: 215–26.

Kumar, M., V. M. Shenbagaraman, R. N. Shaw, and A. Ghosh. 2021. Predictive data analysis for energy management of a smart factory leading to sustainability. *In, Innovations in Electrical and Electronic Engineering* (Springer).

Kumar, P. 2021. Big data analytics: an emerging technology. *In 2021 8th International Conference on Computing for Sustainable Global Development (INDIACom)*, 255–61. IEEE.

Kumar, T. V. V., and M. Haider. 2010. Materialized views selection for answering queries. *In International Conference on Data Engineering and Management*, 44–51. Springer.

L'heureux, A., K. Grolinger, H. F. Elyamany, and M. A.M. Capretz. 2017. Machine learning with big data: Challenges and approaches. *Ieee Access* 5: 7776–97.

Lenzerini, M. 2002. Data integration: A theoretical perspective. *In Proceedings of the twenty-first ACM SIGMOD-SIGACT-SIGART Symposium on Principles of Database Systems*, 233–46.

Menezes, B. C., J. D. Kelly, A. G. Leal, and G. C. Le Roux. 2019. Predictive, prescriptive and detective analytics for smart manufacturing in the information age. *IFAC-PapersOnLine* 52: 568–73.

Naur, P. 1968. 'Datalogy', the science of data and data processes. *In IFIP Congress* (2): 1383–87.

Naur, P. 1974. *Concise survey of computer methods* (Petrocelli Books).

Nida, E. A. 1949. Morphology: The descriptive analysis of words.

Pyle, D. 1999. *Data preparation for data mining* (morgan kaufmann).

Ramírez-Gallego, S., B. Krawczyk, S. García, M. Wo-niak, and F. Herrera. 2017. A survey on data preprocessing for data stream mining: Current status and future directions. *Neurocomputing* 239: 39–57.

Ratner, B. 2017. *Statistical and machine-learning data mining:: Techniques for better predictive modeling and analysis of big data* (CRC Press).

Reddy, D. R. L. C. 2011. A review on data mining from past to the future. *International Journal of Computer Applications* 975: 8887.

Scheuermann, P., J. Shim, and R. Vingralek. 1996. *Watchman: A Data Warehouse Intelligent Cache Manager* (Citeseer).

Sivarajah, U., M. M. Kamal, Z. Irani, and V. Weerakkody. 2017. Critical analysis of Big Data challenges and analytical methods. *Journal of Business Research* 70: 263–86.

Tukey, J. W. 1962. The future of data analysis. *The Annals of Mathematical Statistics* 33: 1–67.

Tukey, J. W. 1977. *Exploratory data analysis* (Reading, Mass.).

Tukey, J. W. 1984. *The collected works of John W. Tukey* (Taylor & Francis).

Yang, Q., and X. Wu. 2006. 10 challenging problems in data mining research. *International Journal of Information Technology & Decision Making* 5: 597–604.

Zhang, S., C. Zhang, and Q. Yang. 2003. Data preparation for data mining. *Applied Artificial Intelligence*, 17: 375–81.

CHAPTER 9

Application of Algorithm on Computational Intelligence and Machine Learning for Product Design
Emerging Needs and Challenges

Sukanta Kumar Baral[1,]* and *Ramesh Chandra Rath*[2]

Contents

[1] Professor, Department of Commerce, Faculty of Commerce & Management, Indira Gandhi National Tribal University (A Central University), Amarkantak, Madhya Pradesh, India.
[2] Professor, Dean (R&D) and HOD MBA, Guru Gobind Singh Educational Institutions Technical Campus, Approved by AICTE Govt. of India New Delhi and Affiliated to Jharkhand University of Technology (JUT) Ranchi, India.
* Corresponding author: drskbinfo@gmail.com

9.1 Introduction

The discovery of enormous records and algorithms as well as recent progress in computing services have led to an unprecedented interest in machine learning. Machine learning techniques are becoming more and more common in large-scale classification, regression, clustering, and dimensionality reduction applications that require high-dimensional input data. Machine-gaining knowledge demonstrates superhuman competencies in plenty of disciplines (self-riding cars, picture classification, etc.). As a result, machine learning algorithms are used in many applications, such as image and speech recognition, internet searches, fraud detection, email and spam filtering, credit ratings, and more. While data-driven research, especially machine learning, has a long history in biology and chemistry, it is relatively new in solid-state materials science.

Artificial intelligence (AI) is a branch of computer science focused on building intelligent machines that behave, function, and respond like humans. AI is used in a real-time context to assist machines in making judgments. An artificially intelligent computer reads real-time data, analyses the business environment, and responds properly. Artificial intelligence devices are developed for a variety of tasks, including such as (a) recognition of speech, (b) learning, (c) planning, and (d) problem-solving.

While there have been several definitions of artificial intelligence (AI) throughout the previous few decades, John McCarthy proposes the following description in this 2004 study: "It is the science and engineering behind creating intelligent devices, particularly clever computer programs." It is akin to the same goal of utilising computers to comprehend human intellect, but AI does not have to limit itself to biologically observable ways. However, the birth of the artificial intelligence debate was marked decades before this definition came about by Alan Turing's seminal work, "Computing Machinery and Intelligence", which was published in 1950.

Artificial intelligence (AI) is a branch of computer science focused on building intelligent machines that behave, function, and respond like humans. AI is used in real-time to assist machines in making judgments. Artificial intelligence (AI) is the replication of human intellect in robots that are programmed to think and act like humans. The term may also refer to any machine that demonstrates human-like characteristics such as learning and problem-solving.

The capacity of artificial intelligence to rationalise and execute actions that have the highest likelihood of reaching a certain objective is its ideal feature. Machine learning is a subset of artificial intelligence that refers to the idea that computer systems can automatically learn from and adapt to new data without the assistance of humans. This autonomous learning is made possible by deep learning techniques.

An artificially intelligent computer reads real-time data, analyses the business environment, and responds properly. Artificial intelligence devices are developed for a variety of tasks, including such as (a) recognition of speech, (b) learning, (c) planning, and (d) problem-solving.

(a) Recognition of Speech:

As far as speech recognition is concerned, it is a device of AI or an interdisciplinary subject of computer science and computational linguistics that develops approaches and technologies that allow computers to recognise and translate spoken language into text, with the primary benefit being search capability through the use of AI algorithms.

(b) Learning:

As far as learning is concerned, it is a process of acquiring new understanding, knowledge, behaviors, skills, values, attitudes, and preferences. The ability to learn is possessed by humans, animals, and some machines, and it may be classified into various types, such as (i) Machine Learning (ML), (ii) Deep Learning (DL), (iii) Artificial Learning (AI).

(c) Planning:

As far as planning in Artificial Intelligence is concerned, it is the process of performing decision-making activities by robots or computer programmes to attain a specified goal. The execution of planning entails selecting a sequence of activities that has a high probability of completing the given assignment.

(d) Problem-solving:

As problem-solving is concerned, it is the process of finding desirable answers to problems relating to a variety of difficulties related to artificial intelligence approaches, including diverse strategies such as building efficient algorithms, heuristics, and doing root cause analysis. The fundamental goal of artificial intelligence is to solve problems in the same way that people do.

Goals of Artificial Intelligence: AI aims to achieve different goals through its realistic Application in different sectors such as:

- To Construct Intelligent and Expert Systems: The purpose of the development was to develop intelligent systems.
- Learning, exhibiting, explaining, and advising its users from various fields are among the tasks that these robots are anticipated to perform.
- Imbuing Machines with Human Intelligence: Creating systems and software that understand, think, learn, and act like humans.

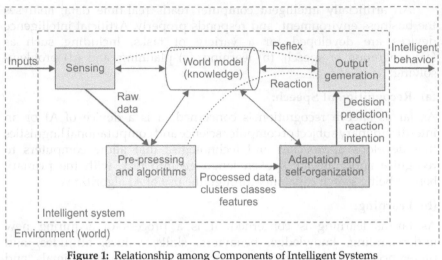

Figure 1: Relationship among Components of Intelligent Systems
Source: https://slideplayer.com/slide/6652373/23/images/69/Relationship%20among%
20components%20Intelligent%20systems.jpg

Contribution of Artificial Intelligence: Artificial intelligence plays a significant role in improving the quality of computer functions such as identification, recognition, simulation of various models, and even production and design processes in fields such as Computer Science, Biology, Psychology, Linguistics, Mathematics and Engineering.

Application of Computational Learning Theory: Computational Learning Theory (CLT) is a well-defined branch of theoretical computer science that uses Machine Learning Algorithms to do Mathematical Analysis. The machine's perception, reaction, and decision-making are all dependent on its capacity to discern various characteristics of the environment using inputs from numerous sensors.

   For example, Computer vision's entire analysis includes face recognition, object recognition, and gesture recognition.

   Another key field that is related to AI is robotics. Robots are capable of a variety of functions, including navigation and object handling. Localization, mapping, and motion planning are the sub-problems.

Programming of Artificial Intelligence: The following examples of AI applications compare the basic programming of a system and show how they differ when designed with and without Artificial Intelligence (AI), such as:

Basic Principles of Machine Learning: In general, machine learning algorithms aim to improve a task's performance by using examples.

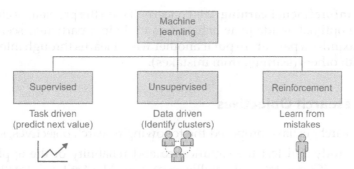

Figure 2: Types of Machine Learning
Source: https://towardsdatascience.com/what-are-the-types-of-machine-learning-e2b9e5d1756f

Machine learning is divided into three types (please refer to Fig. 2): (a) supervised learning, (b) unsupervised learning, and (c) reinforcement learning.

(a) Supervised machine learning: It seeks for an unknown function that relates known inputs to unknown outputs in the same way that a typical fitting approach does. Extrapolation of patterns observed in tagged training data is used to estimate the intended output for unknown domains (task-driven).

(b) Unsupervised learning: It focuses on discovering patterns in unlabelled data, such as clustering samples (data-driven).

Table 1: Refers to the users of AI and CI.

Application	Un use of AI	Use of AI
Artificial Intelligence	Only problems and queries that have already been fed into the system can be solved and answered by the system.	The AI-based system may be used in a variety of settings, analysing data, weighing possibilities, and making judgments.
Computational Intelligence	Any change or addition to the written programme or information can have a significant influence on the structure of the application.	By combining highly independent pieces of data to access a wide variety of facts and make intelligent judgements, AI-enabled programmes, on the other hand, can swiftly adapt to new changes and modifications. As a result, even the slightest modification in the structure of the programme has no effect.
Artificial Intelligence	Contrary to popular belief, changes are not as simple or rapid as they appear. A minor alteration can have a negative impact on the programme, causing it to malfunction.	Modifying AI programmes, on the other hand, is a simple and quick process. These applications are extremely adaptable, and changing changes has no impact on the program's functionality.

(c) **Reinforcement Learning:** It is concerned with the problem of choosing the optimal or adequate behaviours to do in a particular scenario to maximize a payout. To put it another way, it learns through interacting with others (learning from mistakes).

9.2 Research Objectives

The researchers have proposed the following research objectives, such as:

(1) To study and test the significance and reliability of the application of CI (Computational Intelligence) and AI (Artificial Intelligence) on Machine learning when Product design and its manufacturing processes.

(2) To experiment with various applications of AI in product Production, operation and design through the simulation Model.

(3) To examine the reliability and authenticity of the Integrated model how to enhance the computer integrity and assemble the software for smooth identification of the problem and its right solution (Identification, Recognition, and problem Solution).

9.3 Literature Review

In the literature review, most of the data were collected from published and unpublished sources, so both the primary and secondary methods of study were followed by the researchers. Researchers were able to explore the phase and composition space significantly more quickly thanks to the results of Monte Carlo simulations and molecular dynamics studies. Friedler et al. (2016), looked at the definition of a 'fair algorithm', drawing on philosophical and computer scientific perspectives in the future course of research work.

According to AI pioneer John McCarthy, artificial intelligence is defined as "the science and engineering of constructing intelligent machines, especially clever computer programs". As the term indicates, AI is the process of embedding intelligence in robots so that they can act similarly to humans. The approach to computing intelligence known as evolutionary computation (EC) is based on natural evolution. Individuals are then subjected to natural-evolutionary processes such as crossover, mutation, selection, and reproduction. A new population is created using the fitness values of newly developed individuals. The breeding produces a good variety of rabbit genetic material, with some slow rabbits breeding with fast rabbits, some rapid rabbits breeding with fast rabbits, and so on.

AI has become an essential topic of research in all sectors in the twenty-first century, including engineering, science, education, medical, business, accounting, finance, marketing, economics, the stock market, and law, among others. Since the intelligence of computers with machine learning skills has had major effects on corporations, governments, and society, the scope of AI has expanded considerably. They also have an impact on wider trends in global sustainability. Artificial intelligence has the potential to be effective in resolving crucial difficulties in sustainable manufacturing (e.g., optimization of energy resources, logistics, supply chain management, waste management, etc.).

In this regard, there is a movement in smart production to apply AI to green manufacturing processes to strengthen environmental rules. Indeed, as Hendrik Fink, head of Sustainability Services at PricewaterhouseCoopers, stated in March 2019, "If we effectively include artificial intelligence, we may create a revolution in terms of sustainability". "Artificial intelligence will be the driving factor behind the fourth industrial revolution".

As a result, AI subfields such as machine learning, natural language processing, image processing, and data mining have become hot topics for today's tech titans. Because of the constant growth of the technology accessible today, the issue of AI piques the scientific community's attention.

The advancement of ML as a subfield of AI is presently quite rapid. Its use has grown in a variety of sectors, including learning machines, which are now employed in smart manufacturing, medical science, pharmacology, agriculture, archaeology, gaming, and business, among others. Based on the foregoing factors, a comprehensive literature review of AI and the ML approach research from 1999 to 2019 was studied.

Therefore, it is thought important to develop a categorization system that refers to publications that simultaneously discuss the two issues to increase variance and reflection. To acquire a better understanding, the impacts of other variables, such as theme areas and sectors in which the technologies are most important, were also investigated. The key contribution of this chapter is that it gives an overview of previous studies.

9.4 Research Methodology

In this section, researchers have proposed two hypotheses unanimously to justify the aforesaid research objectives for finding out a good outcome-based result.

Null Hypothesis: [H_o]

It refers that an application of computational intelligence has no impact on product design and simulation models of Machine Learning if the manufacturing process, not followed the advanced method of algorithm-based artificial intelligence technology.

Alternative hypothesis: [H$_e$]

It refers to the application of algorithm-based Computational Intelligence on Machine learning that has great importance and is more effective in product production, design and operational function than the nonusers of CI and AI technology.

Models for Principles of Computational and Artificial Intelligence: To establish the model's ability to generalize and extrapolate, it must first be verified with previously unknown data, referred to as the test set. The evaluation can be done in a variety of methods, from a basic holdout to k-fold cross-validation, leave-one-out cross-validation, Monte Carlo cross-validation, and leave-one-cluster-out cross-validation, there's something for everyone. All these decisions during the training phase are based on hiding some variables from the model. The data set is divided into k equal-size sets for the primary validation and the data set is divided into k equal-size sets for the thousandth cross-validation.

In materials science, machine learning is mostly regarded as supervised learning. The volume and quality of data available determine the usefulness of such systems, and this has proven to be one of the key problems in material informatics (Please see Fig. 3 below to understand the principles of computational intelligence).

As a result, appropriately describing data is a critical component of machine learning methodologies in databases such as the Materials

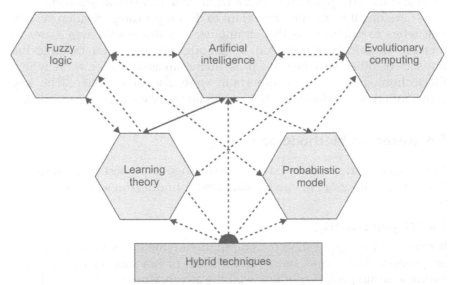

Figure 3: Models for Principles of Computational and Artificial Intelligence.
Source: https://image.slidesharecdn.com/ieeeciapplicationsshared-180107152043/85/
computational-intelligence-and-applications-8-320.jpg?cb=1515338710

Project, the Inorganic Crystal Structure Database, and other facilities. In addition (Genome Materials Initiative) Materials science characteristics must be able to collect all of the essential data to differentiate between various atomic or crystalline environments. The algorithm is mostly in charge of determining the amount of processing necessary. In certain circumstances, such as deep learning, feature extraction can be regarded as a model component.

9.5 Algorithm-Based Application of Computational Intelligence

In supervised machine learning, the researchers started with a projected output and trained the system appropriately. In cases where the outcome is unclear, unsupervised learning is suitable.

Clustering: In other cases, not only will the conclusion be unclear to us, but data characterizing information will also be lacking (data labels). By clustering incoming data pieces that share specific traits, a Machine Learning algorithm can uncover underlying patterns.

When dealing with exceptionally large numbers of variables, clustering can also be utilized to reduce noise (irrelevant factors within the data).

Product Design Simulation: Manufacturers can utilise design simulation to verify and evaluate a product's intended to function as well as its manufacturability. Simulation models are collections of mathematical equations that depict the behaviour of a system in a certain physical area.

Product Simulation Model for Future Optimization: In this section, for justifying the aforesaid research objectives, this model is given for the product simulation and its future optimization, when application of different forms of architecture and their design, there is needs algorithm based on artificial intelligence and computational Intelligence for the

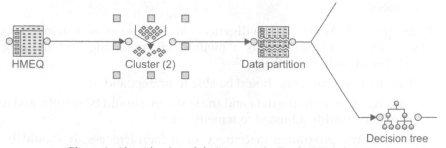

Figure 4: Algorithm-based decision tree for Product Simulation.
Source: https://www.sciencedirect.com/topics/computer-science/modeling-algorithm

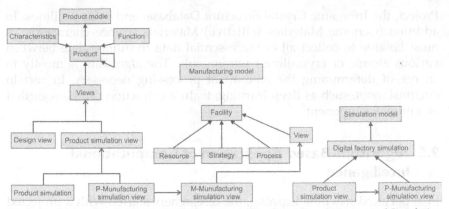

Figure 5: Information Integration between views of Product, Manufacturing and Simulation Models (Ayadi et al. 2011).

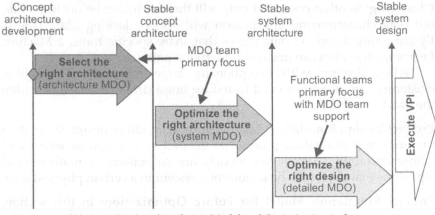

Figure 6. Product Simulation Model and Optimization in future.
Source: https://www.esteco.com/corporate/future-modeling-simulation-and-optimization-cummins-0

optimum function of all product simulation. [See the Fig. 6 Model of simulation]

Techniques of Artificial Intelligence: AI Technique is employed to address these difficulties. It is a method of organizing and utilizing knowledge in such a way that:

• Information providers should be able to understand it:

• Making changes to the data and the software should be simple, and it should be readily adjusted to remedy errors.

• Despite the program's inaccuracy or incompleteness, it should be beneficial in a variety of contexts.

- Given the complexity of AI algorithms, these AI approaches should improve the speed with which these programmes are executed, hence increasing efficiency.

Basic Applications of Artificial Intelligence: In general, AI has been utilized and is dominating in a variety of disciplines where reading and modifying real-time data is required for qualitative research and other tasks, such as:

(a) **Gaming Sector:** In strategic games like Chess, Poker, and Tic-tac-toe, real-time data processing is essential. The system should be able to weigh a variety of choices before making a judgment based on heuristic knowledge. In these strategic games, artificial intelligence (AI) is critical.

(b) **Natural Language Processing:** Machines must be able to grasp the language of numerous users for software to function successfully. Not only should the system be able to adapt to many languages, but also dialects and accents. In a variety of situations, AI has proven to be extremely effective.

(c) **Expert Systems:** An intelligent machine's primary function is to make decisions. These devices require software that takes data as input, interprets it, weighs numerous choices, and decides. These computers are used to provide a rationale for the circumstance at hand. Users can make informed judgments with the help of software that provides explanations and suggestions.

(d) **Vision Systems:** It delivers visual input, which is a vital and difficult-to-understand type of information. As a result, before making decisions based on visual inputs, a system with intelligence must read, understand, analyse, and comprehend it.

(e) **Production Sector:** In this industry, a well-designed and innovative product will be achievable thanks to the deployment of artificial intelligence software, which will gather resources and perform integrated functions using ML, DL, and SL.

1. **Speech Recognition:** Some AI-enabled systems are built to be able to hear and interpret spoken language to understand what is being said. When a human speaks with the system in many languages, the system not only understands the words, but also the sentences, their meanings, and the tone. Accents, dialects, slang words, background noise, variances in speech modulation, changes in voice owing to discomfort, cold, and other variables are all recognized by the computer.

2. **Handwriting Recognition:** Text-reading software of this type has been developed. This text can be written with a pen or pencil on paper. A

mouse or a stylus can be used to write the text on a computer screen. It can read the text and recognize letter and number shapes before transforming it into editable text that may be modified, stored, and amended.

3. **Intelligent Robots:** Robots are machines that have been programmed as slaves to carry out a master's commands. They have a variety of sensors on board. Physical data from the real environment is collected by these sensors.

Performance Table of Product Simulation Model

For testing different models of simulation on the application of algorithm-based artificial intelligence and CI on machine learning the following performance will be reflected in the following data table and semiotic model of the Bar chart as given below.

Table 2: Data Table.

Name of Models	Mode of Simulation	Algorithm based Mutation with AI and CI		% of Performance	Result
		Yes	No		
Product Manufacturing Simulation Model (PMSM)	Machine Learning	yes		98.30	Highly Performed
Probabilistic Model (PM)	AI and CI		No	82.00	Average Performed
Hybrid Technic Model (HTM)	Machine Learning	Yes		99.10	Very highly
Algorithm based Decision Tree Model (ABDTM)	Machine Learning	Yes		98.86	Highly

Hypothesis Testing: From the above-said data table and result table, researchers have come to give a justified hypothesis testing on the observation of ourselves in various application and mutation processes of CI and AI on Machine learning and finally proves that the Application of Algorithm based on CI and AI on Machine learning has great importance and effect on assembling of resources, for product production and design while, manufacturer goes for integrate function. Besides that, this software almost uses full in computer education, integration, identification and recognition and problem-solving skills etc. Therefore, due to sufficient proof against the Null hypothesis (Ho) has not true because algorithm-

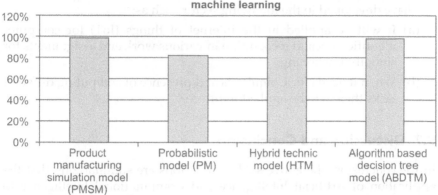

Figure 7: Semiotic Model of the simulation model in bar-Chart.

Table 3: Final result.

Model and its category	Mode of Simulation	Percentage of performance	Performance in Rank
Product Manufacturing Simulation Model (PMSM)	Machine Learning	98.30	3rd
Probabilistic Model (PM)	AI and CI	82.00	4th
Hybrid Technic Model (HTM)	Machine Learning	99.10	1st
Algorithm based Decision Tree Model (ABDTM)	Machine Learning	98.86	2nd

based AI and CI are highly necessary used in Machine Learning. Therefore, the Null hypothesis has been rejected and accept the alternate hypotheses (H_e).

9.6 Findings

From the above discussion, the researchers have identified the following findings to recommend to the research world for facilitating research generation in future. Such as:

- Advanced technology for the application of Artificial Intelligence (AI) and Computational Intelligence (CI) would be advanced in Hybrid systems as well as new technologies and algorithms will be applied when the production process and Product design.

- The new application and uses of Computational Intelligence would have developed in the following sector such as:
  - (a) It will be applied in the internet of things (IoT) for computer identification and recognition in various work and assignments for smooth functioning.
  - (b) When it needs for Ubiquitous and presence of computing different tasks there will be applied in future.

9.7 Discussion and Conclusion

In the conclusion, the above-said summations are summarizing that the application of Artificial Intelligence and Computational Intelligence to Algorithm-based Machine Learning, is an excellent effect on Product design and its functions are tremendously controlled and managed. An algorithm based on Artificial Intelligence (AI) has several benefits, but it also has several disadvantages and limitations. While manufacturers concentrate on product development and design, it is up to the company to decide whether such investments are necessary and profitable.

9.8 Limitations and Future Scope

Despite the many merits, algorithm-based artificial intelligence on machine learning; has some limitations such as:

- For architecture, building, rebuilding, and repairing a structure involves the employment of competent professionals and is costly in terms of both money and time because it is expensive to store things.
- Memory access and retrieval may be slower than in humans.
- Machines can be programmed to learn and develop over time, but they will never be as good as humans, and their operations are confined to the software they wrote.
- They could never have the same level of creativity as humans.
- Because of advancements in intelligent machines, unemployment is the greatest concern.
- Humans, being lazy, can become overly reliant on technology and underutilize their cerebral powers, so the Application of AI and CI is so useful in future courses of research work, new product design and Production.
- Machines can easily lead to destruction in the wrong hands, but when new technology-based AI and CI execute in the manufacturing of products, operating, of machines it will be cent performed.

References

Ayadi, M., R. C. Affonso, V. Cheutet, F. Masmoudi, A. Riviere, and M. Haddar. 2011. Proposition d'un modèle d'intégration des simulations pour l'Usine Numérique. 7ième Conférence Internationale Conception & Production Intégrées CPI'11, Oujda, Maroc.

Baldi, P., and S. Brunak. 2001. Bioinformatics: The machine learning approach. The MIT Press, Cambridge, MA.

Chan, P. K., and S. J. Stolfo. 1998. Toward scalable learning with non-uniform class and cost distributions: a case study in credit card fraud detection. pp. 164–168. *In:* Agrawal, R., Stolorz, P., and Piatetsky, G. (eds.). *KDD'98 Proc. Fourth International Conference on Knowledge Discovery and Data Mining.* AAAI Press, New York.

Friedler, S. A., C. Scheidegger, and S. Venkata Subramanian. 2016. Vol.abs/1609.07236, CoRR (CORR 2016), 2016.

Guzella, T. S., and W. M. Caminhas. 2009. A review of machine learning approaches to spam filtering. *Expert Syst. Appl.* 36: 10206–10222.

Hohenberg, P., and W. Kohn. 1964. Inhomogeneous electron gas. *Phys. Rev.* 136: 864–871.

Huang, C.-L., M.-U. Chen and C.-J. Wang. 2007. Credit scoring with a data mining approach based on support vector machines. *Expert Syst. Appl.* 33: 847–856.

Pazzani, M. and D. Billsus. 1997. Learning and revising user profiles: the identification of interesting websites. *Mach. Learn*, pp. 313–331.

Prompramote, S., Y. Chen, and Y. P. P. 2005. Machine learning in bioinformatics. *In:* Chen, Y. P. P. (eds.). *Bioinformatics Technologies.* Springer, Berlin, Heidelberg (2005). https://doi.org/10.1007/3-540-26888-X_5.

Rajan, K. 2005. Materials informatics. *Mater Today* 8: 38–45.

Swamidass, S. J., C. Azencott, H. Gramajo, S. Tsai, and P. Baldi. 2009. The influence relevance voter: an accurate and interpretable virtual high throughput screening method. *Journal of Chemical Information and Modeling* 49(4): 756–766.

Silver, D., A. Huang, C. J. Maddison, A. Guez, Laurent Sifre, G. van den Driessche, J. Schrittwieser, I. Antonoglou, V. Pennershevlam, M. Lanctot, S. Dieleman, D. Grewe, J. Nham, N. Kalchbrenner, I. Sutskever, T. Lillicrap, M. Leach, K. Kavukcuoglu, T. Graepel and D. Hassabis. 2016. Mastering the game of go with deep neural networks and tree search. *Nature* 529: 484–489.

Section III
Business Intelligence and Analytics Applications

CHAPTER 10

HR ANALYTICS
Galvanizing the Organizations with the Prowess of Technology

Ruchi Jain and *Ruchi Khandelwal**

Contents

10.1 Introduction
10.2 Literature Review
10.3 Applications of HR Analytics
10.4 Research Methodology
10.5 Data Analysis and Interpretation
10.6 Challenges in the Successful Implementation of HR Analytics
10.7 Conclusion

10.1 Introduction

The digitalization of human resource management has uncovered more inventive and innovative approaches. It has led the organization to opt for competency-based recruitment, newer methods of compensation, assessment of outcomes for performance, etc., for empowering the human resources. The people have always been the assets in an organization and an advantage to sustain in the competitive environment. So, progressive organizations must endorse recent emerging trends while

Amity School of Business, Amity University Uttar Pradesh, India.
* Corresponding author: rkhandelwal@amity.edu

maintaining their best human resource. This requires a lot of planning, organizing, directing, staffing, and controlling while making any decision. Technological advancement can be of great help in the prevailing dynamic environment and to handle employee behavior efficiently. Moreover, the digitalization of work processes with the aid of new technologies can transform the workplace culture and will make it more professional and objective in orientation. The arrival of big data-driven approaches is replacing employee evaluation to assess their commitment to the organization as compared to existing methods of research surveys and reviews by specialists (Kumar 2020). Organizations are significantly using data science to collect, analyze and investigate information about a candidate or employee from all the digital touchpoints like social media platforms, opinion polls and virtual communities (Kampakis 2020).

So, the HR department also deals with a great volume of valuable data about the company and its employees in HR information systems. This database holds promise to make a rational decision with the help of digitalization tools and data analytics in planning, recruiting, and managing human resources. HR analytics is the emerging discipline adopted by organizations to give preference to analytics at all stages ranging from a candidate getting on board, to performance monitoring and stretching to scrutinizing the social media opinions of the employees.

Traditionally, the process of HR analytics starts by collecting the data of employees from HRIS (Human Resource Information System), from previous performance records, and from any interface of the organization and social media. This data produces a data warehouse where data mining techniques are administered to understand the concealed data patterns or probabilities for anticipating the following trends:

- Forecasting the demand and supply of employees,
- Identifying the relevant test for recruitment as per job profile,
- Reviewing the training needs as per the new skill set required,
- Performance appraisal and maintaining relevant information for it,
- To decide the incentives based on performance,
- Managing the overall profile of the employee.

Thus, HR analytics helps HR managers to perform their functions and duties to contribute toward organizational growth effectively and efficiently. HR analytics is also referred to as People Analytics Talent Analytics or Workforce Analytics. The research objectives of the chapter are as follows:

To study the modus operandi of usage of HR Analytics in an organization.

To study the companies practising HR analytics and explore the variety of reasons for it.

To find the relation between employee performance and Motivation with the usage of HR Analytics tools.

10.2 Literature Review

Analytics enable HR to make strategic contributions, but not all analytics offer equal insights. According to Levenson 2005, "HR analytics includes the use of statistical techniques, research designs and algorithms to evaluate employee data and translate results into evocative reports." He compared the usefulness of ROI, cost-benefit, and impact analysis and emphasized that it is the right time for HR to build an HR analytics centre of expertise and create a foundation of analytic skills across the function.

Mondore et al. (2011) propagated that HR analytics created an opportunity to show the direct impact of HR processes on business outcomes. Though the definitions and process details associated with HR analytics have not been well-articulated in the desired manner. Smith (2013) stated that HR analytics is an area in the field of analytics that refers to applying analytic processes to the human resource department of an organization in the hope of improving employee performance and therefore getting a better return on investment. HR analytics does not just deal with gathering data on employee efficiency. Instead, it aims to provide insight into each process by gathering data and then using it to make relevant decisions about how to improve these processes.

Usage metrics and HR analytics were also studied by Lawler et al. (2004) in corporations. Their results indicated that HR functions often collect data on their own efficiency. However, it never collects data on the business impacts of its programs and practices. This is a crucial point because those HR organizations that collect effective data are more likely to be strategic partners. This finding suggests that if HR wants to play a strategic role in organizations it needs to develop its ability to measure how human capital decisions affect the business and how business decisions affect human capital.

According to Watson 2010, Narula 2015, there are three categories of analytics:

A. **Descriptive analytics:** when a process involves understanding the historical data, predicting behavior, and getting outcomes by describing the relationship, it is referred to as the first level of analytics. It majorly involves visualization of data, report formation, and describing relationships. Example: turnover rates, absence rates, cost per hire.

B. **Predictive analytics:** It is the second level of analytics which involves decision making, pattern recognition, forecasting, predictive analysis, and data mining. It focuses on the correlation between data.

C. **Optimization analytics:** It helps in finding the best alternative training investment to reach the organizational goal by using linear programming or creating mathematical modelling by using limited resources. This is the third level of analysis. Analytics help HR managers forecast and perform their functions in a more logical and detailed way. This will save costs and help in optimizing resources. This keeps the organization and its employees at pace with the dynamic and competitive external environment.

Barbar et al. (2019) tested four hypotheses to establish a relation between HR analytics and the development of employee skills and retention of employees. The results revealed that HR managers rely on HR analytics to formulate employee development strategies. Mohammed (2019) explored the existing literature in the field of HR analytics and their implication for predictive decision-making in an organization. Reviewing the literature on the integration of HR analytics in organizational setups and the introduction of relevant IT infrastructure and provisions will be critically included. Some components, such as multi-levelled equations, graphics, and tables are not prescribed, although the various table text styles are provided. The formatter will need to create these components, incorporating the applicable criteria that follow.

Extant literature indicates HR analytics as an exhibit of HR metrics put in an arranged manner. Bassi (2011) established it as a business solution which is logically based on 'Predictive Modelling' and 'What-if scenarios' hence increasing the significance and credibility of the organization. Belizón and Kieran (2021) viewed HR analytics as a procedure of data-driven decision-making by the HR department. It addresses the strategic and operational level issues of HR by making analytical usage of data by the routine process of:

1. Research design for the HR concern
2. Data management for decision making
3. Data analysis for problem-solving
4. Data interpretation and communication
5. Designing subsequent action plans and
6. Evaluation of plan and learning.

Opatha (2021) agreed that HR analytics is used for sustainable decision-making as a part of organizational strategy with the help of the accuracy of big data. However, he critically argues that data-driven decisions from HR analytics may not be the best solutions being highly automated. These decisions need to be moderated further as per the financial condition, environmental variables, competitiveness, and competencies of the organization.

10.3 Applications of HR Analytics

10.3.1 Talent Acquisition

The key to success for any company lies in its skill pool and therefore companies struggle hard to entice the most capable professionals.

Through the use of data science and modern-day software companies can sieve through thousands of resumes and generate a group of the most talented prospects, from where applicants can be taken for the next round of interviews. With the help of big data analytics, the process becomes more defined and proficient.

Juniper Networks, a networking and cyber security solutions Company, uses its big data to analyze not only where the top-performing employees come from but also where they go after leaving the company. Juniper Networks aims to gain insight into distinct industry career paths and develop new and contemporary strategies for attracting and retaining current and future talent.

Black Hills Corporation, a 130-year-old energy conglomerate, expanded its workforce to about 2,000 employees after acquisition and was faced with the challenges like an ageing workforce, the necessity for specialized skills, and an extended timeline for training employees to make them fully competent. The company used workforce analytics to calculate how many employees would retire per year, the types of workers needed to replace them, and where those new hires were most likely to come from.

Google uses an effective hiring algorithm for predicting which candidates had the highest probability of succeeding after they are hired.

10.3.2 Training and Development

Training is essential not just for recruits but also for existing employees to ensure that they keep upgrading themselves as industry processes and work methods change. Companies are undertaking huge investments to ensure employees attend online courses, training sessions, and other learning programs. However, it is difficult to assess the precise benefits of such training and many companies realize that the cost-benefit analysis of training is not positive. With the ability of data science, it is possible to study employee learning and scrutinize the cost-benefit analysis for the various training programs that are conducted and modify courses to make them more productive.

Ameriprise Financial is a diversified financial services company demerged from American Express in 2005. The newly formed firm's HR activities such as onboarding, training and performance reviews were rated poorly by the employees in the beginning. Furthermore, Ameriprise Financial lacked an effective plan for allocating HR time to talent issues with the highest impact. To improve their HR functions, Ameriprise combined workforce and financial data to align talent investments with business results, and more proactively develop data-driven insights used to predict turnover, reduce new hire failure rates, and manage persistent poor performers.

One of the most fruitful ways to successfully integrate the data with organizational development is through the 70–20–10 rule, which is used for determining the perfect balance between providing staff development and corporate learning opportunities. **SAP** has been effectively using the 70–20–10 rule for many years in different ways in its several departments for evolving specific learning programs. Combining HR analytics with this rule, the company has made on-the-job, social, and formal learning a vital part of its training process. Moreover, the 70–20–10 rule also has its roots in the company's key processes like employees' annual development plans.

10.3.3 Employee Performance

Analysing employee performance is one of the most significant contributions of big data in HR. The system enables a company to observe all key performance indicators in real-time and to appraise each one of its employees individually. Big data analysis also reveals probable faults and flaws in work, as valued feedback to employees that can be used to take a corrective course of action quickly.

A revolutionary initiative at **Google** is PiLab, a unique subgroup that conducts applied experiments within the company to determine the most efficient methods for managing people and maintaining a productive environment. It even predicts the type of reward that makes employees the happiest. As a result, they have many engaged employees with an average participation rate of 90%, ultimately showcasing their success in improving business methods and morale.

Another case in point is the logistics giant, **UPS,** which has provided its drivers with intelligent handheld computers that support drivers in making informed decisions, such as determining the order of delivering parcels for the most efficient route. Additionally, the company collects essential statistics on the driver's behavior through the usage of around

200 sensors that are fitted onto the trucks. These sensors record data on most slight actions of the driver such as wearing or the number of times the vehicle was reversed. This data is then used to provide feedback to the drivers and suggest improvements or training as needed. The insights have had a major impact on the financials of the company reducing consumption of 8.5 million gallons of fuel and covering 85 million miles per year. The company drivers can make more deliveries per day with an average of 120 stops a day as opposed to less than 100 in the past.

10.3.4 Employee Compensation

Using advanced analytics, the HR teams in companies can determine the finest financial model for each employee of the company. Compensation management solutions help organizations visualize all their compensation data in one place. HR departments use these solutions to streamline compensation planning and reporting. Having this data in a single dashboard also empowers leaders to view current compensation policies and develop fair compensation strategies for their teams.

Clarks a British-based, international shoe manufacturer and retailer, founded in 1825, has over 1,000 branded stores and franchises around the world. The company applied compensation and benefits analysis to optimize rewards packages for employees. By asking which benefits employees might be prepared to trade off, it built a much more granular view of what people truly valued and adjusted the compensation package accordingly.

10.3.5 Employee Retention

Retention is the top priority for businesses all over the world and hardly there would be any organization in any sector not struggling with keeping the employees intact.

As early as 2015, **Credit Suisse** used innovative churn analytics to predict who might quit the company. The company was able to predict not just who might quit, but also *why* these people might quit. This information was provided anonymously to managers so they could reduce turnover risk factors and retain their people better. In addition, special managers were trained to retain high-performing employees who had a high turnover risk. In total, the company saved approximately $ 70,000,000 a year as a result of this practice.

Google developed a mathematical algorithm to proactively and successfully predict which employees are most likely to become a retention problem.

Bank of America experienced a turnover problem with its call centres leading to poor customer experience and customer frustration. The company leveraged data collected from its call centers using analytics to understand the root cause of such a high turnover rate. Using the insight generated, the bank improved its business policies finally resulting in the company saving $15 million with the increased productivity and decreased employee turnover.

10.3.6 Employee Engagement

According to Tim Young, (CMO, Yellobrick Data, UK), "Employee engagement is paramount to attracting and retaining talent, and remaining competitive in the global market. Research shows a direct link between employee engagement and corporate performance". Research indicates that employee productivity is directly affected by employee happiness (American Business Magazine 2013). Job dissatisfaction is a common phenomenon troubling employees, be it work-life balance, poor working conditions, low pay or lack of career advancement plans, employees quote several reasons for their growing dissatisfaction.

Google has been using analytics to adapt different aspects of its people processes to fit its unique work culture. One such notable initiative was Project Oxygen; a crucial discovery was made that great managers do lead to more engaged and productive teams. Google researched further and building on research findings, they developed Google's top 8 management behaviours that are used to train and identify the most effective leaders in the company.

10.4 Research Methodology

This research is a descriptive research study where HR Analytics tools and techniques are studied to find a greater relationship between employee performance and Herzberg's Two Factors of motivation. For conducting this research, a company is chosen where HR Analytics can be studied-Anupama printing solutions Pvt. Ltd. From the pool of working population of the company, 25 random people were selected as a representative sample to fill the questionnaire and their responses are recorded for study through Random sampling. The employees were given a structured questionnaire where they had to rate each factor. For attaining the research objectives, MS-Excel has been used to find the correlation coefficient to determine the relationship strength as statistical tools and techniques used.

10.4.1 Brief of the Sample Company

Anupama Printing Solutions Pvt. Ltd is based in Delhi, NCR. It was set up in 2009 and was the largest prepress unit in the Delhi NCR region. The company business deals with a solution for newspaper printing, posters, banners, book printing, etc. To enhance efficiency, the company installed BAYS PRINT 581 a CTP machine, to expose the offset plates used for printing. Earlier their daily production was three hundred plates per day which increased to six hundred plates. On average the daily production grew by seventy per cent and the clients were enjoying their fast services. Their main aim is to deliver service timely and without any failure. By 2013, they started night shift production which turned the company into a 24/7 business. In the last few years, they have diversified further towards printing and labelling. There is a backup machine to ensure that the work never stops. The company is led by it's director Mr. Tarun Katyal.

10.4.2 A Brief About the Research Tool (Theoretical Basis of Herzberg's Theory)

Herzberg's Two Factor Theory or the motivator-hygiene theory was developed by Frederick Herzberg in 1959. The research was conducted by asking a group of people to share their experiences at work, good and bad. Depending on the responses he developed two kinds of factors. Factors for satisfaction (motivator) and factor for dissatisfaction (hygiene factors) as depicted in Fig. 1. Motivators are those which increase the satisfaction of employees and the hygiene factor can't motivate but the absence of these factors can lead to high dissatisfaction among employees.

Taking this theory as the basis of research, the factors (hygiene and motivators) are identified, and they are inserted in the questionnaire to obtain the response of the employees of the sample company. The following factors (Table 1) are considered for the study.

A structured questionnaire was administered among the employees of Anupama Printing solutions and the employees were asked to rate each factor on a five-point Likert's scale.

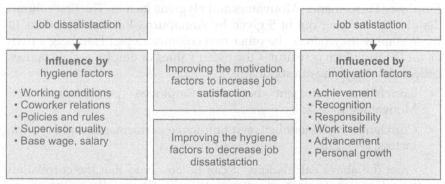

Figure 1: Herzberg's Two Factor Theory or the motivator-hygiene theory.
Source: https://www.toolshero.com/psychology/theories-of-motivation/two-factor-theory-herzberg/

Table 1: Mapping of Herzberg's theory factors with research motivational Variables.

	Hygiene factors	
	Factors in Herzberg theory	**Variables mapped (in study)**
1.	Working conditions	Working environment is comfortable
2.	Co-workers' relations	Employees are treated with due respect
3.	Supervisor quality	Supervisor is cooperating
4.	Policies and rules	Communication system is good
5.	Policies and rules	Good work done is appreciated
6..	Base wage and salary	Salary structure
	Motivational factors	
1.	Achievement	Performance based incentives
2.	Recognition	Promotion opportunities
3.	Responsibilities	Job rotation and new assignments
4.	Work itself	Training and development programme
5.	Advancement	Worker's participation in management activities
6..	Personal growth	Reimbursement for official work

10.5 Data Analysis and Interpretation

The employee performance of Anupama Printing Solutions Pvt Ltd was measured by rating their performance on a 5-point scale. In this scale of ranking 5 is considered as excellent performance, 4 as a good performance, 3 as an average performance, 2 as a below average performance and 1 as a bad performance. The performance measured is the average performance of employees in the last six months that is collected by the company. Table 2 below depicts the data of 25 samples in the heads of Employee Performance, Motivators and Hygiene factors. The first column denotes the ranking out of 5 given by Annapurna Printing Solutions to these employees whereas the other two columns depict Herzberg's two-factor theory which constitutes the mean values of employees' responses collected through the questionnaire.

✓ Correlation coefficient between Employee performance and Motivators: R= 0.305.

✓ Correlation coefficient between Employee performance and Hygiene factors: R= 0.228.

From the above calculations, it can be observed that the correlation coefficient of employee performance to motivators falls in the range of +0.30 to +0.39, i.e., 0.305 which means the relationship strength is a moderate positive relationship.

Table 2: Representative data from the research survey.

S. No.	Employee performance	Motivators	Hygiene factors
1	4	2.17	2.83
2	3	2.83	2.50
3	4	4.00	2.67
4	3	1.33	1.33
5	3	2.17	3.50
6	3	2.17	3.17
7	2	2.67	2.67
8	2	1.83	2.83
9	2	2.17	2.33
10	4	2.83	1.83
11	4	3.00	2.67
12	5	3.33	3.17
13	2	2.33	2.50
14	3	2.00	2.67
15	3	3.00	3.00
16	4	3.33	3.17
17	2	2.83	2.83
18	4	4.33	4.17
19	3	3.67	3.00
20	2	4.83	4.00
21	2	4.00	2.50
22	3	2.83	3.33
23	4	3.83	3.67
24	3	3.33	3.33
25	5	4.33	3.67

Whereas the correlation coefficient of employee performance to hygiene factors lies in the range of +0.20 to +0.29 i.e., 0.228 which means the relationship strength is a weak positive relationship.

Pearson's correlation coefficient is used to measure the strength of the relationship between two variables. Here it can be observed that there is a greater impact of "Motivators" on employee performance. The factors mentioned by Herzberg in Motivators are

Ø Performance-based incentives

Ø Promotion opportunities

Ø Job rotation and new assignments

Ø Training and development programs

Ø Worker's participation in management activities

Ø Reimbursement for official work

From the data collected it can be interpreted as Motivational factors that the majority of the employees tend towards the Training and Development programme in the Motivator factor. Only 15% of employees are motivated by reimbursement for official work. This shows that the employees in the organization want to learn and develop their knowledge and skills. Since it's a manufacturing company, training programmes can help the employees to be updated with the new technology and can decrease the rate of defects. Whereas in the Hygiene Factors the outcome was that most of the employees have given an almost equal preference to Salary structure, appreciation, relationship with their supervisor and equal treatment in the organization. This shows that people are concerned about development and recognition. They want a fair working environment where the supervisors are cooperating and appreciate their good work.

All these factors should be considered while taking decisions related to employee performance. The organization Anupama Printing Solutions Pvt Ltd can investigate these factors to motivate their employee, while performance appraisal, employee retention etc process their decision making.

From this study, it can be observed that the correlation between employee performance and motivators is higher than the hygiene factors. So, the company can take up these factors and focus on them to improve their employee performance. The company can hire HR Analytics to provide them with such logical or optimized solutions, which can contribute to their growth and success. The study has used one such tool which is correlation and made the stored data into data visualization. Here descriptive analysis which is the first level of analysis is done to identify some relationship between different factors. Here it can be observed that HR Analytics helps the HR manager to perform such functions in a more optimized and effective way.

10.6 Challenges in Successful Implementation of HR Analytics

It is not just that HR functions are complex, but the metrics developed for performance through HR analytics are also not easy to implement (Opatha 2020). Fernandez and Gallardo (2020) identified fourteen different

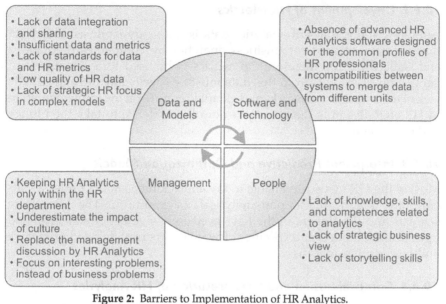

Figure 2: Barriers to Implementation of HR Analytics.
(Derived from *Fernandez, V., and Gallardo-Gallardo, E. (2020). Tackling the HR digitalization challenge: key factors and barriers to HR analytics adoption. Competitiveness Review: An International Business Journal.*

barriers that make the adoption of HR Analytics difficult for companies. They further grouped these factors into four categories, namely data and models, software and technology, people, and management.

While the first three categories discuss the needed resources for HR Analytics implementation, the last category considers the strategic framework needed to support the implementation process. Figure 2 gives a clear picture of these fourteen different factors.

10.6.1 Organizing Quality Data

It is fundamental to curate and integrate data, gathered from different processes and operations and divisions within the company to implement HR analytics. The data must be made available in a manner that it can be assessed sincerely, however, the quality of data, pulling data from varied sources and converting data into meaningful insights remains a big challenge (Hota 2021). HR analytics today is integrated cross-functionally with other departments for the achievement of business objectives. But the functional silos in which organizations work make data accessibility a key concern to deal with.

10.6.2 Development of HR Metrics

Another issue is that HR analytics data is not as dynamic as other data that is being utilized for business analytics by companies (Fernandez and Gallardo 2020), therefore the metrics to be used for evaluation and generation of insights from this data must be specifically developed. The lack of theoretical frameworks and extant literature on the implementation of HR analytics puts the managers out of confidence to take the plunge and experiment.

10.6.3 Insufficient Predictive and Optimization Models

Most of the HR metrics available today are lacking as many professionals currently are from a non-analytical background. The practising scholars and developers in the field need to evolve better statistical and optimization models and specific software that permit detailed predictive and prescriptive analysis.

10.6.4 Comprehension and Interpretation of HR Analytics

The usage of analytics produces multifaceted models and reports, interpretation of which is relatively very difficult (Davenport and Anderson 2019). The complex machine learning tools are subject to doubt as the executives find it difficult to explain the causes of their success. Thus, the manager's skills to interpret and use the data must be upgraded.

10.6.5 Division of Opinion about the Onus of HR Analytics

Although HR analytics is about people, it is not to be restricted to HR professionals alone. Value-added analytics can develop only through the collaboration of functional experts in all domains (Rasmussen and Ulrich 2015, McIver et al. 2018). This shall lead to giving a strategic perspective to the analytics function.

10.6.6 Leveraging the Company's Resources

One of the keys to building lasting competence for an organization is its ability to leverage resources for generating needed capabilities. Similarly, to digitize HR processes, it is vital to leverage the existing firm's resources. Translating new job positions and finding the right people, managing employee attrition and compensation issues can be tackled to give outstanding outcomes once leveraged properly (Hota 2021).

10.6.7 Privacy Laws and Data Abuse

Data is the lifeblood of HR analytics and so maintaining data privacy is essential to maintain sanctity in the organization among the employees. Companies need strict data protection policies and systems in place to prevent data theft and abuse. But at the same time, the HR analytics team may find it difficult to process the data within the bounds of legislation and data-protection regulations (Van Den Heuvel and Bondarouk 2017).

10.6.8 Translation of Insights into Action

The analytics function may develop crucial insights into both HR and non-HR-related domains. These understandings must be integrated into the planning and implementation framework of the firm to come up with meaningful plans of action. But there can be procedural delays and a lack of interpersonal understanding that may prevent this (Dykes 2016).

10.7 Conclusion

In organizations, HR managers want to address key strategic business issues. They are aware of the growing technology and the need for competent human resources for the survival of the organization in the market. Thus, HR analytics is a crucial and essential practice to develop skills and provides the link between the HR policies & practices the organizational performance. HR analytics can produce comprehensive reports using the personnel data stored in the organization. It can provide insights and predictions that can be used by HR managers to develop their HR strategies. In other words, they can be termed KPIs 2.0. Worldwide, companies are using it for applications like talent acquisition, Training and development, assessing employee performance and compensation along with ensuring employee engagement for employee retention. Companies at the forefront like Google are using HR analytics to create a unique organizational culture for a highly motivated workforce. Implementing HR analytics can be challenging also as the data available may be inadequate to develop the requisite HR matrices and make the translation of insights into action more difficult for the organization. Thus, HR Analytics can produce advanced operational quantitative reports, strategic outputs, predictive analysis, and wider operational reports, for completely creating data visualization for the predictive applications of HR for empowered organizations.

References

Angrave, D., A. Charlwood, I. Kirkpatrick, M. Lawrence, and M. Stuart. 2016. HR and analytics: why HR are set to fail the big data challenge. *Human Resource Management Journal* 26(1): 1–11.

Ansho, K. 2021. HR analytics–empowering organizations. *Artificial Intelligence and Green Thinking*, p.80.

Ballinger, G. A., R. Cross, and B. C. Holtom. 2016. The right friends in the right places: Understanding network structure as a predictor of voluntary turnover. *Journal of Applied Psychology*, 101(4): 535.

Barbar, K., R. Choughri, and M. Soubjaki. 2019. The impact of HR analytics on the training and development strategy-private sector case study in Lebanon. *Journal of Management and Strategy* 10(3): 27.

Bassi, L. 2011. Raging debates in HR analytics. *People and Strategy* 34(2): 14.

Belizón, M. J., and S. Kieran. 2021. Human resources analytics: A legitimacy process. *Human Resource Management Journal*.

Carter, L. 2020. HR Data and Analytics: How it Can Make a World of Difference. Available via https://blog.bestpracticeinstitute.org/hr-data-analytics-can-make-world-difference/. Accessed 31 January 2022.

Cook, I. 2021. HR Moments Matter: 3 Compensation Challenges Solved With People Analytics. Available via https://www.visier.com/blog/culture/compensation-people-analytics/. Accessed 11 Febraury 2022

Davenport, T., and D. Anderson. 2019. *HR Moves Boldly into Advanced Analytics with Collaboration from Finance*. Technical report, ORACLE.

Dykes, B. 2016. Actionable insights: The missing link between data and business value. *Retrieved December* 1: 2017.

Fenech, R. 2022. Human resource management in a digital Era through the lens of next generation human resource managers. *Journal of Management Information & Decision Sciences* 25.

Fernandez, V., and E. Gallardo-Gallardo. 2020. Tackling the HR digitalization challenge: key factors and barriers to HR analytics adoption. *Competitiveness Review: An International Business Journal*.

Fineman, D. R. 2017. People analytics: Recalculating the route 2017 Global Human Capital Trends. Available via https://www2.deloitte.com/us/en/insights/focus/human-capital-trends/2017/people-analytics-in-hr.html. Accessed 12 Febraury 2022.

Giermindl, L. M., F. Strich, O. Christ, U. Leicht-Deobald, and A. Redzepi. 2021. The dark sides of people analytics: reviewing the perils for organisations and employees. *European Journal of Information Systems*, pp. 1–26.

Hladio. M. 2013. Happy Bosses + Happy Employees = More Profits. Available via http://www.americanbusinessmag.com/2013/10/happy-bosses-plus-happy-employees-equal-more-profits/. Accessed 11 February 2022.

Hota, J. 2021. Framework of challenges affecting adoption of people analytics in india using ism and micmac analysis. *Vision*, p.09722629211029007.

Kampakis, S. 2020. Hiring and managing data scientists. pp. 105–123. *In The Decision Maker's Handbook to Data Science*. Apress, Berkeley, CA.

Kumar, M. B. 2020. Applications of Data Science in HR Analytics. Available via https://360digitmg.com/applications-of-data-science-in-hr-analytics. Accessed 8 November 2021.

Lawler III, E. E., A. Levenson, and J. W. Boudreau. 2004. HR metrics and analytics–uses and impacts. *Human Resource Planning Journal* 27(4): 27–35.

Marler, J. H., and J. W. Boudreau. 2017. An evidence-based review of HR Analytics. *The International Journal of Human Resource Management*, 28(1): 3–26.

McIver, D., M. L. Lengnick-Hall, and C. A. Lengnick-Hall. 2018. A strategic approach to workforce analytics: Integrating science and agility. *Business Horizons*, 61(3): 397–407.

Mohammed, D., and A. Quddus. 2019. HR analytics: A modern tool in HR for predictive decision making. *Journal of Management* 6(3).

Mondore, S., S. Douthitt, and M. Carson. 2011. Maximizing the impact and effectiveness of HR analytics to drive business outcomes. *People and Strategy* 34(2): 20.

Narula, S. 2015. HR analytics: Its use, techniques and impact. *CLEAR International Journal of Research in Commerce & Management* 6(8).

Neill, M. 2020. How 5 Successful Companies Are Using HR Analytics. Available via. Accessed 11 Febraury 2022.

Opatha, H. H. D. P. J. 2020. HR Analytics: A literature review and new conceptual model. *International Journal of Scientific and Research Publications* 10(6): 130–141.

Opatha, H. H. D. P. J. 2021. HR Analytics: A Critical Review-Developing a Model Towards the Question "Can Organizations Solely Depend on HR Big Data Driven Conclusions in Making HR Strategic Decisions all the Time?".

Rasmussen, T., and D. Ulrich. 2015. Learning from practice: how HR analytics avoids being a management fad. *Organizational Dynamics* 44(3): 236–242.

Reddy, P. R., and P. Lakshmikeerthi. 2017. HR analytics-An effective evidence-based HRM tool. *International Journal of Business and Management Invention* 6(7): 23–34.

Sheriff, S. 2019. Advanced Analytics In HR: Applications And Examples. Available via https://acuvate.com/blog/advanced-analytics-in-hr-applications-and-examples/. Accessed 11 February 2022.

Smith, T. 2013. *HR Analytics: the what, why and how*. Numerical Insights LLC.

Tembhekar, P. 2021. Human Resource Information Systems: Implementing Data Analytics Techniques in Human Resource Functions.

Van den Heuvel, S., and T. Bondarouk. 2017. The rise (and fall?) of HR analytics: A study into the future application, value, structure, and system support. *Journal of Organizational Effectiveness: People and Performance*.

Watson, H. J. 2010. BI-based Organizations. *Business Intelligence Journal* 15(2): 4–6.

Yeh, T. Y., and M. Raheem. 2021. In-depth Analysis of Extant Business Analytics: A Review from the Human Resource Perspective.

CHAPTER 11

Marketing Analytics
Concept, Applications, Opportunities, and Challenges Ahead

Manita Matharu

Contents

11.1 Introduction

Regardless of the industry in which they operate, businesses employ a variety of marketing analytics. Technology-driven techniques and solutions are critical for optimizing marketers' connections with clients in a large and unpredictable market environment. By focusing on interaction and feedback, marketing and customer analytics have now become industry keywords as they help marketers to gain a more in-depth understanding of their consumers' experiences and make appropriate changes as a result. Analytics is used in a wide variety of fields, including consumer

Amity School of Business, Amity University Uttar Pradesh, India.
Email: manita.mktg@gmail.com

electronics, intelligent objects, and a rising number of other applications. The ability to make better and faster decisions based on data is becoming increasingly important for marketing firms. Marketers must adapt and evolve their analytic strategies, capabilities, and data solutions to remain competitive in an environment where new technologies and channels are being introduced daily. This data must be gathered, confirmed, and transformed into useful information before it can be used. Then after, the information must be examined, interpreted, and translated into actionable intelligence. Subsequently, the intelligence must be completely related to the goals and requirements of the organisation for it to be actionable. Marketing gets more credible and usable as a result of following this process to the fullest.

Marketing analytics has seen a surge in popularity in recent years, owing to the benefits it can provide to businesses. Several organisations are now cognizant of the fact that marketing analytics can provide them with a substantial competitive advantage. Organizations that have developed a mature analytics culture are better able to use data in their business decisions. Numerous organisational functional areas increasingly rely on data and analytics to increase their knowledge and influence (Ransbotham and Kiron 2019). Analytics has progressed from an afterthought to a demand in a data-driven and automated society (Logi Analytics 2018). Several businesses have thrived because of their capacity to collect, examine, and act on data. The expertise of these organisations can benefit any industry (Davenport 2006). To survive and flourish, intelligence must be consistently integrated further into customer experience, offerings, processes, and services (Logi Analytics 2018). Customer segmentation is critical for any industry's marketing of products and services. By tailoring marketing and product offerings to specific customers based on their purchasing behaviour, you can boost profitability by increasing customer response rates. Market segmentation is implemented in the project through the use of algorithms found in Big Data analytics and other related fields.

Businesses all across the world are increasingly incorporating sustainability issues into their business planning. The Sustainable Development Goals (SDGs) of the United Nations emphasize the critical role that a sustainable lifestyle will play in achieving the expected paradigm shift toward sustainability in the future. However, while the environmental literature recommends profiling, segmentation must place a greater emphasis on psychographics than on more conventional measures such as demographic features (Straughan and Roberts 1999). Without a doubt, the lifestyle choices of customers influence their purchasing behaviour. Nonetheless, little research has been done on the influence of changing consumer lifestyles, specifically the impact of consumers' concerns for sustainability on their purchase decisions in developing nations, on their purchasing decisions. The purpose of this study:

(a) To provide a comprehensive overview of "marketing analytics"— concept and evolution.
(b) To explore the application areas of marketing analytics.
(c) To examine Indian consumers' existing lifestyle dimensions in terms of health and sustainability, as well as to identify the lifestyle characteristics that these consumers possess.

This chapter begins the evolution of marketing analytics followed by some of the existing definitions, highlighting their necessity and importance. Following that, the application areas of marketing analytics are covered, as well as the opportunities and challenges associated with marketing analytics and the conclusion.

11.2 Literature Review

11.2.1 Evolution of Marketing Analysis

The concept, characterization, and explanation of marketing as a science have been around for nearly a century. Marketing analytics can be traced back to the late 1950s and early 1960s when numerous market research businesses made their household panel data available to academic academics (Sheth 2021). Marketing data was accessible at the aggregate level, on a yearly or monthly basis, and was therefore difficult to analyse. Nielsen developed one of the first and most well-known market research organisations to track product sales in retail establishments in 1923, making it one of the first and most well-known firms in the world. The field of traditional analytics matured in the mid-1990s, Internet marketing became popular, and marketers realised the value of using log files to track website visitor interactions. CRM software from companies like Oracle became available at the same time (Wedel and Kannan 2016). I/PRO Corp was the first commercial web analytics company to launch in 1994, followed by WebTrends in 1995, Omniture in 2002, and Google Analytics in 2005 (Chafey and Patron 2012). The new analytics have even had an impact on marketing research, allowing researchers to use web-based interactive survey tools, online qualitative analysis, database mining, and analysis (Hauser 2007). In the words of Seth 2021, "Starting in the sixties, marketing began to diverge into three distinct subdisciplines: Marketing Strategy, Consumer Behavior, and Marketing Analytics. Each sub-discipline offered its courses and its textbooks. Finally, each sub-discipline has its membership associations (AMA, ACR, and INFORMS/ISMS) with their academic journals (Journal of Marketing, Journal of Marketing Research, Journal of Consumer Behavior and Marketing Science)".

11.2.2 Marketing Analytics—The Concept

Marketing analytics is getting more popular among various business enterprises in the digitalized corporate environment of the twenty-first century. Marketing analytics is the act of evaluating, controlling, and analysing market performance to maximise overall effectiveness and investment return (www.wordstream.com). Since the turn of the millennium, there have been numerous significant research endeavours aimed at identifying the notion of marketing analytics, each with its analytical base (Davenport and Harris 2017). The researchers concentrated their attention on the sense that could be used to bridge the gap between academic and practitioner comprehension. Marketers have defined marketing analytics as drawing attention to the descriptive, diagnostic, predictive, and prescriptive stages of marketing science for informative data reservoirs, for functional connectedness of marketing science in the modern world to sustain a competitive advantage and achieve better outcomes through smarter decisions as a result (Davenport 2006, Farris et al. 2010, Davenport and Harris 2017).

According to Krishen and Petrescu (2018), marketing analytics has an impact on many aspects of an organisation that includes "decision-making, marketing activities, management, management information systems (MIS), operations management, finance, and economics". Marketing analytics is a "technology-enabled and model-supported approach to harness customer and market data to enhance marketing decision making" (Germann et al. 2013, Lilien 2011: p. 5).

According to the "Global Marketing Analytics Market Report 2021 Featuring market Leaders—Adobe, IBM., Microsoft., SAS Institute, Teradata, and Wipro", Market analytics is described as the process of identifying important trends in data to make more informed marketing decisions. Businesses in a variety of industries use marketing analytics to make data-driven decisions about how to spend their marketing budget most effectively. Businesses must be equipped with the necessary measurement tools to execute an efficient and lucrative marketing strategy. Businesses can use marketing analytics to precisely determine the efficacy of their campaigns and make data-driven decisions. Rather than basing marketing initiatives on fragmented data sources or assumptions, marketing analytics helps firms to see the whole picture across all marketing channels, enabling them to make informed marketing decisions (Research and Markets 2021).

According to marketingevolution.com, marketing analytics examines market research data to find patterns in things like campaign conversions, consumer behaviour, regional preferences, creative preferences, and more. Marketing analytics is used to optimise future campaigns based on prior successes (marketingevolution.com 2022). Marketingevolution.com suggests that initiating the marketing analytics process involves four steps. The first is, to determine what needs to be measured first. Conversion rates, lead generation, and brand recognition may all be quantified based on the data or problem that needs to be solved or understood. Next is to establish a standard for what constitutes an effective campaign. This will shape the data and metrics marketers collect. An increase in brand loyalty assessed by a customer panel, instead of an internet click, maybe the success metric for raising brand awareness. The next step, is to assess current capabilities, such as what is the company doing now. What are your flaws? Evaluating offline advertising results or choosing the most convertible medium can help you improve your programme. Lastly, utilize marketing analytics methods. Customer selection and data proliferation will make marketing analytics solutions important.

11.2.3 Marketing Analytics—Applications

Analytically mature business entities, according to Ransbotham and Kiron (2019) make effective usage of data and information from many sources. According to Bruce Bedford, Vice President of Oberweis Dairy Inc.'s marketing analytics and consumer insights department, marketing analytics can help a brand advance the quality of its customer engagement with its target market (Ransbotham and Kiron 2019). In "Big Data, Marketing Analytics, and Public Policy: Implications for Health Care," Kopalle and Lehmann (2021) offer enlightening directions. Their perspective, in collaboration with Chen et al. (2021), outlines how analytics can be used to mix technology and data to derive marketing insights.

Although, Davis et al. 2021 introduce six areas for future research in marketing analytics and public policy, that includes retail analytics; social media analytics; marketing-mix analytics. According to Krishen and Petrescu (2018), Marketing analytics has a substantial impact on a variety of organisational processes, including decision-making, operations management, finance, and economics. Figure 1 depicts the various business disciplines that marketing analytics has an impact on.

Today's marketing needs more precise analytics than ever. Sponsored media has become increasingly unpopular among consumers. In today's tech-driven environment, firms use marketing data to better interact with customers (Samala et al. 2019). Social media competitive analytics is increasingly being used by organisations to better understand the complete customer experience, according to He et al. (2016). By concentrating on

Figure 1: Marketing analytics in business activities.
Source: (Krishen and Petrescu 2018)

interaction and feedback, customer analytics helps marketers better understand their consumers' experiences. Chatting with consumers, polling certain demographics, or engaging them in surveys can help organisations obtain a deeper grasp of their goods' differentiators and competitive advantages. Teams may therefore more precisely match things as per individual consumer interests. Analytics can also shed light on the kinds of product features that customers want. Marketing departments may share this information with product advancements for future releases. Additionally, data analysis can help marketers decide the best locations to display messages to certain consumers. While conventional marketing channels like print, television, and radio continue to be used, marketers must understand which digital platforms and social media platforms customers prefer. Analytics also implies that users see more targeted, personalised ads that cater to their specific interests, rather than more annoying bulk advertising (marketingevolution.com 2022).

11.2.4 Analytics for Market Segmentation—Consumer Lifestyle Segmentation

Analytics may disclose a lot about clients. For instance, what kind of message/creativity works best with them? Which products do they buy and which do they search for? Which ads are clicked and which are overlooked? Analytics also aids in streamlining or improving the buyer's experience. Marketing analytics contribute to trend forecasting, efficiency improvement, fraud detection, and true business innovation (Ahmed et al. 2017). In today's global environment, predictive analytics is critical for marketing professionals to gain a better understanding of their customer's demands and purchasing patterns, allowing them to position their products more accurately. Businesses are increasingly planning for sustainability. Indeed, the Sustainable Development Goals (SDGs) acknowledge the importance of a sustainable lifestyle in attaining future sustainability. The purpose of this study is to envision and comprehend consumer trends and preferences based on their lifestyle choices in the context of sustainability. This study categorises customers based on their lifestyle choices using clustering techniques.

In the mid-1990s, Ray and Anderson (2001) proposed the acronym LOHAS. Because LOHAS consumers represent a fast-expanding worldwide market, they are becoming increasingly valuable to a wide range of businesses. NMI (2008) categorizes the market into five segments: LOHAS, naturalists, drifters, ordinary, and indifferent. The LOHAS group is said to be very environmentally conscious, with high levels of education, dependable marital status, and high income. The LOHAS group also values health, prefers organic foods, conserves natural resources like water and power, and prefers fabric bags over plastic bags. LOHAS stands for healthy and eco-friendly products and consumers. Experts claim that LOHAS is a significant and rapidly growing subset of consumer research (Wenzel 2007, Paulesich 2008). Without a doubt, LOHAS represents a multi-billion-dollar market opportunity and businesses are eager to extend their LOHAS product offers (Oppermann 2008, Häyrinen et al. 2016). Still, to the researcher's knowledge, no study has been done on Indian customers' LOHAS propensity. Thus, the current study attempts to incorporate sustainability into market segmentation by examining Delhi NCR consumers' LOHAS preferences. This study attempts to categorize customers in the Delhi NCR based on their health and sustainability-oriented lifestyles (LOHAS) and to make recommendations to customers regarding sustainable product marketing and promotion.

11.3 Research Methodology

This study identifies a significant segment of customers interested in a healthy and sustainable lifestyle in India's Delhi-NCR region. The study attempts to segment customers according to their lifestyle choices. Due to the difficulty that an ignorant person has in comprehending and responding to environmental and sustainability-related concerns, the survey data was gathered online from customers of Delhi and the National Capital Region of India (Kumar et al. 2017, Vermeir and Verbeke 2006). Several things (such as 'SwachhBharat' and the use of green and blue dumpsters) were rewritten to match Indian customers' language quality and altered in the Indian context. Finally, 820 people completed the survey.

The instrument was created to ascertain consumers' predisposition to LOHAS. The 12 assertions were created using a document analysis technique that used measuring scales adapted from past research on LOHAS which includes Yeh and Chen (2011), Lehota et al. (2013), NMI (Natural Marketing Institute), Lee (2005), and Kwak (2006). This study employed modified statements that were embedded for use in India. On a five-point Likert scale, each item was rated (5 being "strongly agreed" and 1 being "strongly disagreed").

11.4 Analysis and Findings

First, the 12 characteristics that define consumer lifestyles were factored in. After five iterations, the Varimax rotation compressed the 12 variables into three factors that account for 68.36 per cent of the variation (Table 1). The KMO sufficiency score (0.827) is high, as is Bartlett's test of sphericity (5266.479), which is statistically significant at 0.000. For clustering, the rotated component matrix organises the elements in decreasing order of explanatory power.

The three retrieved components explain 68.367 per cent of the variation. The factor loading ranged from 0.634 to 0.920. Support for the Clean India campaign, waste separation (blue bins/green bins), and recyclable packaging were used to assess the first component, 'Environment friendliness.' It was .822 reliable. Consumers who protect the environment and charitable work, provide safe working conditions for their employees, support "Make in India", and are willing to pay more for organic vegetables and fruits were determined by respondents' responses to five questions, with reliability of =.850. 3rd element: health consciousness is based on replies to prior questions concerning activity, healthy eating preferences, and belief in healthy food suggested by friends and/or media. Overall, the 12 items were 0.861 reliable. Table 1 includes the factors, their items, and their factor loadings.

Following factor analysis, these variables were used as inputs in subsequent research. This study used a mix of hierarchical and non-hierarchical clustering to determine segmentation. Although the popular combination of speed and hierarchical clustering methods can be imprecise due to outliers (Hair et al. 1998), non-hierarchical clustering methods are less influenced by outliers. As a result, combining hierarchical and non-hierarchical clustering techniques has been recommended (Reynolds et al. 2002). Three to four groups were chosen for research using hierarchical clustering. To classify the customer groupings, the factor scores were used in a k-means cluster analysis. The final clusters were identified using k-means cluster analysis, with the best statistical conclusion being a three-cluster solution. The findings of cluster analysis are shown in Table 2.

Table 1: Consumer lifestyle factors and reliabilities.

Factors	Items	Components			% of Variance	% of Cumm. Variance
		F1	F3	F3		
Environment Consciousness	I am active in helping the Swachh Bharat Mission in environmental protection		.700			
	I carefully segregate the waste (blue bins/green bins)		.836			
	It is crucial to recycle the product.		.816		26.793	26.793
	I prefer things that are created sustainably (organic farming, nature conservation products)		.686			
	Cronbach's alpha		.822			
Values supporting sustainability	I like companies that share value with healthy and sustainable lifestyle consumers that protect the environment	.634				
	I consistently support donations in the activities of charities.	.780				
	I believe that the company that offers its employees with good working circumstances is a superior organisation.	.800			21.596	48.389
	I focus on economic prosperity of the country by purchasing Indian products that promote 'India.'	.821				
	I'm willing to pay a premium for natural goods such as organic veggies and fruit.	.759				
	Cronbach's alpha	.850				

Table 1 contd. ...

...Table 1 contd.

Factors	Items	Components			% of Variance	% of Cumm. Variance
		F1	F3	F3		
Health Consciousness	I usually workout to maintain my fitness and health.			.804		
	I choose foods that help me stay healthy..			.920	19.978	68.367
	I agree in friends and the media referring to healthy eating options.			.897		
	Cronbach's alpha **Overall scale reliability**			*.864* **.861**		

Extraction Method: Principal Component Analysis.
Rotation Method: Varimax with Kaiser Normalization.
a. Rotation converged in 5 iterations.

Table 2: Cluster Analysis—results.

	Cluster 1		Cluster 2		Cluster 3	
	(Consumer's orientation towards Healthy Lifestyle)		(The Indifferent)		(Consumer's orientation towards Lifestyle of Health and Sustainability)	
Variables	Mean	SD	Mean	SD	Mean	SD
Values supporting sustainability	–.829	.553	–.270	.530	.964	.760
Environment Consciousness	–.328	.965	–.003	.908	.301	1.00
Health Consciousness	.813	.397	–1.32	.572	.299	.470
Cluster size	287		247		316	

Source: Primary data

Orientation towards a healthy lifestyle was assigned to 287 respondents (34%) who scored low on sustainability and environmental stewardship principles but high on health consciousness, earning them the title of "The Indifferent". With 316 responses (37%) in the third category, it is clear that this segment of customers is also concerned about and respecting sustainability. This category is titled "Healthy and Sustainable Lifestyle Orientation".

ANOVA was used to examine whether there are any significant differences in the means of clusters and to validate the determined categories. The cluster groups differ significantly in terms of environmental responsibility, sustainability values, and health consciousness, as demonstrated in Table 3. We have no idea where such disparities exist because the ANOVA test reported in Table 3 indicated that the findings are significant in aggregate. Tukey's HSD is used to determine the significance of ANOVA results. Tukey's test is used to assess whether two groups' means are statistically different. Tukey's post hoc test is used to determine if the groups depicted in Table 4 differ significantly.

The study's goal was to divide the market based on lifestyle characteristics including health and sustainability, allowing business leaders to adopt policies and plans that embrace sustainability. After establishing the study's most important elements, segments were produced using a mix of hierarchical and non-hierarchical clustering approaches. Findings indicated the clusters are identified by meanings linked with the three categories (health consciousness, sustainability values, and environmental stewardship). Inclusion in that cluster is based on the variables having the highest positive or negative counts.

Segment 1—Consumer's orientation toward a healthy lifestyle (34 per cent)
This market is dominated by young consumers (18–25 years old) who are health sensitive yet lack values supporting sustainability and environmental friendliness. In this group, males (55%) outweigh females (44 per cent). This cluster has the most 'Single' residents (59 per cent).

Table 3: Analysis of variance ANOVA.

		Sum of Squares	df	Mean Square	F	Sig.
Values supporting sustainability	Between Groups	509.888	2	254.944	636.773	.000
	Within Groups	339.112	847	.400		
	Total	849.000	849			
Environment Consciousness	Between Groups	59.708	2	29.854	32.036	.000
	Within Groups	789.292	847	.932		
	Total	849.000	849			
Health Consciousness	Between Groups	653.490	2	326.745	1415.540	.000
	Within Groups	195.510	847	.231		

Source: Primary data

Table 4: Results of cluster analysis: Multiple Comparisons.

Tukey HSD

Dependent Variables	(I) Cluster Number of Case	(J) Cluster Number of Case	Mean Difference (I-J)	Std. Error	Sig.	95% Confidence Interval	
						Lower Bound	Upper Bound
Values supporting sustainability	1	2	-.55969025*	.05491756	.000	-.6886276	-.4307529
		3	-1.79476838*	.05159459	.000	-1.9159040	-1.6736328
	2	1	.55969025*	.05491756	.000	.4307529	.6886276
		3	-1.23507813*	.05373932	.000	-1.3612492	-1.1089070
	3	1	1.79476838*	.05159459	.000	1.6736328	1.9159040
		2	1.23507813*	.05373932	.000	1.1089070	1.3612492
Environment Consciousness	1	2	-.32477507*	.08378349	.000	-.5214849	-.1280652
		3	-.63004322*	.07871390	.000	-.8148505	-.4452359
	2	1	.32477507*	.08378349	.000	.1280652	.5214849
		3	-.30526815*	.08198595	.001	-.4977577	-.1127786
	3	1	.63004322*	.07871390	.000	.4452359	.8148505
		2	.30526815*	.08198595	.001	.1127786	.4977577
Health Consciousness	1	2	2.14088980*	.04169889	.000	2.0429877	2.2387919
		3	.51395850*	.03917576	.000	.4219803	.6059367
	2	1	-2.14088980*	.04169889	.000	-2.2387919	-2.0429877
		3	-1.62693130*	.04080425	.000	-1.7227330	-1.5311296
	3	1	-.51395850*	.03917576	.000	-.6059367	-.4219803
		2	1.62693130*	.04080425	.000	1.5311296	1.7227330

Source: Primary data

* The mean difference is significant at the 0.05 level.

Segment 2—'The Indifferent' (29 per cent)
This cluster is dominated by men (62%) whereas the third is dominated by women (65%). (37 per cent). This portion was dubbed the "Indifferent" segment since it showed the least concern for health and the environment. Compared to other clusters, this one has the highest share of recent graduates (49%) and the lowest (15%) of postgraduates, valuing education and knowledge in sustainability, having most singles (55%) than married (44 per cent).

Segment 3—Consumer's orientation to LOHAS (Lifestyle of Health and Sustainability) (37 per cent)
The third group of clients, termed "The Pioneers of LOHAS", is highly health-conscious and environmentally conscious. This group is dominated by women (57%). This category has a higher proportion of married people (58%) than single people (45 per cent). The age bracket 26–35 is the most popular (34 per cent).

11.5 Marketing Analytics—Opportunities and Challenges

Utilizing marketing analytics can increase revenue, reduce costs, improve real-time response rates, and improve decision-making efficiency by converting large amounts of sensor-collected data into valuable business knowledge (Ahmed et al. 2017, Weinberg et al. 2015). Targeted offerings and personalised services benefit both consumers and marketers, resulting in a win-win situation for all parties involved in the system. For example, data analytics helps enterprises to offer multifaceted mechanisms or to link and share information across their organisations (Ahmed et al. 2017, Iacobucci et al. 2019). Market analytics' long-term viability is depending on how technology improves and changes in the coming years. The development and implementation of advanced analytics methodologies are possible in the future, with a significant impact on the growth and overall profitability of large organisations.

Given the fact that marketing analytics are crucial to the success of campaigns, the progression of analytics is fraught with difficulties as a result of the vast volumes of data that marketers may now collect. To gain actionable insights, marketers must determine the most effective approach to organising data to maximise their effectiveness. Among the most critical marketing analytics challenges that marketer face today are the following: Firstly, the quantity of data as the digital era ushered in the era of big data, which enabled marketing teams to track every click and view by the consumer. This volume of data nevertheless is hollow unless it is categorised and evaluated for insights that enable in-campaign adjustments. Accordingly, organizations are interested in the most effective approach to organising data to determine its meaning. Marketers collect data from several sources, and they must find a way to normalise it for it to be comparable. It is particularly difficult to relate online and offline connections since they are usually measured using distinct attribution models, which makes comparison difficult. Not only is there a problem with the massive amount of data that organisations must sort through, but this data is widely regarded as unreliable. Secondly, it is the quality of data. To enable employees to make informed decisions, organisations require a mechanism for ensuring data quality. Aside from that, data scientists are in short supply; even if firms have access to the necessary data, many do not have access to appropriate professionals.

11.6 Conclusion

Successful marketing campaigns in the twenty-first century demand a holistic approach to the entire process, from customer acquisition to lifetime value. Despite its relative infancy in the marketing industry, market analytics has the power to transform many elements of marketing. It is the result of rapidly evolving technology and analytics. In other regards, in the future, truly innovative market analytics may be offered to assist marketers in optimising vital company operations and enhancing business firm profitability (Germann et al. 2014). Therefore, there is a need for academic marketing departments must incorporate marketing analytics as a fundamental focus of their curriculum. Students must not only learn analytical tools and how to apply them, but they must also understand why this process is a vital component of the marketing discipline. Analysis and, particularly, interpretation, must be incorporated into the basic curriculum of marketing departments (Hauser and Lewison 2005). This should be more than a quick introduction to business statistics. At the very least, marketing research courses must contain analytics on an equal footing with quantitative and qualitative research. Not only will this necessitate a shift in how most students perceive marketing, but it will also necessitate professors to understand the processes and change long-held pedagogical practices. Now, marketing analytics has evolved into a useful tool for marketers. As new technology and procedures become accessible shortly, this value will continue to rise.

Integrated marketing analytics programmes are a critical component of this strategy. It is therefore a matter of connecting this data to the company's strategic business context. The present study employed cluster analysis to determine which customer groups are more inclined to buy sustainable products. The study identified three distinct customer segments based on health, environmental, and sustainability preferences:

"Health Conscious", "The Indifferent", and "The Pioneers of LOHAS". This innovative market segmentation strategy allows marketers to better address consumer sustainability perspectives. Identifying separate customer sustainability categories can help organisations establish marketing strategies and discover pioneers or advocates for the adoption of sustainable products in India. The sampling method and self-reporting questionnaire employed in this study limit the generalizability of the findings. To better understand sustainable customer behaviour and utilise sustainable marketing approaches, this study should help. This study has certain drawbacks. So the survey mainly includes Delhi-NCR consumers. Second, the data are from a convenience sample and hence do not accurately represent all ages. The study needs a larger sample size. Future studies can examine how Indian customers choose brands, products, and retailers.

References

Ahmed, U., N. Khalid, A. Ahmed, and M. H. Shah. 2017. Assessing moderation of employee engagement on the relationship between work discretion, job clarity and business performance in the banking sector of Pakistan. *Asian Economic and Financial Review* 7(12): 1197–1210.

Bradlow, T. Eric M. Gangwar, P. Kopalle, and S. Voleti. 2017. The role of big data and predictive analytics in retailing. *Journal of Retailing* 93(1): 79–95.

Chaffey, D., and M. Patron. 2012. From web analytics to digital marketing optimization: Increasing the commercial value of digital analytics. *Journal of Direct, Data and Digital Marketing Practice* 14(1): 30–45.

Chen, Y. T., E. W. Sun, M. F. Chang, and Y. B. Lin. 2021. Pragmatic real-time logistics management with traffic IoT infrastructure: Big data predictive analytics of freight travel time for Logistics 4.0. *International Journal of Production Economics* 238: 108157.

Davis, B., D. Grewal, and S. Hamilton. 2021. The future of marketing analytics and public policy. *Journal of Public Policy & Marketing* 40(4): 447–452.

Davenport, T., and J. Harris. 2017. Competing on analytics: Updated, with a new introduction: The new science of winning. *Harvard Business Press.*

Davenport, T. H. 2006. Competing on analytics. *Harv. Bus. Rev.* 84: 98–107.

Germann, F., G. L. Lilien, and A. Rangaswamy. 2013. Performance implications of deploying marketing analytics. *International Journal of Research in Marketing* 30(2): 114–128.

Germann, F., G. L. Lilien, L. Fiedler, and M. Kraus. 2014. Do retailers benefit from deploying customer analytics? *Journal of Retailing* 90(4): 587–593. doi:10.1016/j.jretai.2014.08.002.

Hair, J. F., W. C. Black, B. J. Babin, R. E. Anderson, and R. L. Tatham. 1998. Multivariate data analysis. Uppersaddle River. *Multivariate Data Analysis (5th ed.) Upper Saddle River* 5(3): 207–219.

Häyrinen, L., O. Mattila, S. Berghäll, and A. Toppinen. 2016. Lifestyle of health and sustainability of forest owners as an indicator of multiple use of forests. *Forest Policy and Economics* 67: 10–19.

Hauser, W. J. 2007. Marketing analytics: The evolution of marketing research in the twenty-first century. *Direct Marketing: An International Journal* 1(1): 38–54.

Hauser, W. J., and D. M. Lewison. 2005. Creating the Comprehensive direct interactive Marketing program. *Journal for Advancement of Marketing Education–Volume* 6: 1.

He, W., X. Tian, Y. Chen, and D. Chong. 2016. Actionable social media competitive analytics for understanding customer experiences. *Journal of Computer Information Systems* 56(2): 145–155.

Kappelman, L., E. McLean, V. Johnson, and N. Gerhart. 2014. The 2014 SIM IT key issues and trends study. *MIS Quarterly Executive* 13(4): 237–263.

Kopalle, P. K., and D. R. Lehmann. 2021. Big Data, marketing analytics, and public policy: Implications for health care. *Journal of Public Policy & Marketing* 40(4): 453–456.

Krishen, A. S., and M. Petrescu. 2018. Marketing analytics: Delineating the field while welcoming crossover. *Journal of Marketing Analytics* 6(4): 117–119. doi:10.1057/s41270-018-0046-6.

Kumar, V., J. B. Choi, and M. Greene. 2017. Synergistic effects of social media and traditional marketing on brand sales: Capturing the time-varying effects. *Journal of the Academy of marketing Science* 45(2): 268–288.

Kwak, Y. S., M. S. Hwang, S. C. Kim, C. S. Kim, J. H. Do, and C. K. Park. 2006. A growth inhibition effect of saponin from red ginseng on some pathogenic microorganisms. *Journal of Ginseng Research*, 30(3): 128–131.

Lee, J. H. 2005. A study on the defecation pattern and lifestyle factors of female high school and college students in Gyeonggi province. *Korean Journal of Community Nutrition* 10(1): 36–45.

Lehota, J., Á., Horváth, and G. Rácz. 2013. A potenciális LOHAS fogyasztók megjelenése Magyarországon. *Marketing & Menedzsment* 47(4): 36–54.

Lilien, G. L. 2011. Bridging the academic–practitioner divide in marketing decision models. *Journal of Marketing* 75(4): 196–210.

Logi Analytics. 2018 State of Embedded Analytics. The Sixth Annual Review of Embedded Analytics Trends and Tactics; Logi Analytics: McLean, VA, USA, 2018; Available online: https://goo.gl/BL7euZ (accessed on 28 September 2019).

Oppermann, M. 2008. Digital storytelling and American studies: Critical trajectories from the emotional to the epistemological. *Arts and Humanities in Higher Education* 7(2): 171–187.

Paulesich, R. 2008. Sustainable consumption: A crucial aspect of research in economics. *Progress in Industrial Ecology, an International Journal* 5(1-2): 149–159.

Samala, N., S. Singh, R. Nukhu, and M. Khetarpal. 2019. Investigating the role of participation and customer-engagement with tourism brands (CETB) on social media. *Academy of Marketing Studies Journal* 23(1): 1–16.

Sheth, J. 2021. New areas of research in marketing strategy, consumer behavior, and marketing analytics: the future is bright. *Journal of Marketing Theory and Practice* 29(1): 3–12.

Straughan, R. D., and J. A. Roberts. 1999. Environmental segmentation alternatives: A look at green consumer behavior in the new millennium. *Journal of Consumer Marketing*.

Ransbotham, S., and D. Kiron. 2017. Analytics as a Source of Business Innovation. The Increased Ability to Innovate is Producing a Surge of Benefits across Industries. 2017. Available online: http://ilp.mit.edu/media/news_articles/smr/2017/58380.pdf (accessed on 2 March 2020).

Ransbotham, S., and D. Kiron. 2019. Using Analytics to Improve Customer Engagement. Sloan Business Review. Retrieved from https://sloanreview.mit.edu/projects/using-analytics-to-improve-customer-engagement/.

Ray, P. H., and S. R. Anderson. 2001. The cultural creatives: How 50 million people are changing the world. Broadway Books.

Research and Markets, 2021, https://www.researchandmarkets.com/reports/5367882/marketing-analytics-2021-2026?utm_source=GNOM&utm_medium=PressRelease&utm_code=5x98p9&utm_campaign=1568933+-+Global+Marketing+Analytics+Market+Report+2021+Featuring+market+Leaders+-+Adobe%2c+IBM.%2c+Microsoft.%2c+SAS+Institute%2c+Teradata%2c+and+Wipro&utm_

exec=chdo54prd.https://www.marketingevolution.com/knowledge-center/the-role-of-marketing-analytics-in-identifying-creative-needs?hsLang=en.

Reynolds, K. E., J. Ganesh, and M. Luckett. 2002. Traditional malls vs. factory outlets: Comparing shopper typologies and implications for retail strategy. *Journal of Business Research* 55(9): 687–696.

Vermeir, I., and W. Verbeke. 2006. Impact of values, involvement and perceptions on consumer attitudes and intentions towards sustainable consumption. *Journal of Agricultural and Environmental Ethics* 19(2).

Wedel, M., and P. K. Kannan. 2016. Marketing analytics for data-rich environments. *Journal of Marketing* 80(6): 97–121.

Weinberg, B. D., G. R. Milne, Y. G. Andonova, and F. M. Hajjat. 2015. Internet of Things: Convenience vs. privacy and secrecy. *Business Horizons* 58(6): 615–624.

Wenzel, P. 2007. Public-sector transformation in South Africa: Getting the basics right. *Progress in Development Studies* 7(1): 47–64.

Wordstream. (n.d.). Marketing Analytics-Success Through Analysis. Retrieved from, https://www.wordstream.com/marketing-analytics/.

Yeh, N. C., and Y. J. Chen. 2011. On the everyday life information behavior of LOHAS consumers: A perspective of lifestyle. *Journal of Educational Media and Library Science* 48(4): 489–510.

CHAPTER 12

Effect of Social Media Usage on Anxiety During a Pandemic
An Analytical Study on Young Adults

Amit Dangi,[1],* Vijay Singh[2] and Neha Gupta[3]

Contents

12.1 Introduction

In the 21st century, the involvement of social media in everyone's life is increasing exponentially. Social media is an active source for information, as during a pandemic the circulation of the newspaper is reduced many folds, and at the same time the online traffic increases at a phenomenal rate, The Hindu has seen a 15 to 20 per cent increase in the online traffic

[1] Faculty of Commerce and Management, SGT University, Gurugram.
[2] Department of Commerce, Indira Gandhi University, Meerpur, Rewari, India.
[3] Amity School of Business, Amity University, Uttar Pradesh, Noida.
* Corresponding author: amit.dangi@sgtuniversity.org

(World Association of Newspapers and News Publishers). Being on the web is not fashion now; instead, it is a necessity. As per Statista, the number of online users in India is 564.5 million in 2021 and is expected to rise to 666.4 million by 2023. Out of these, social network users will increase from 376.1 million in 2021 to 447.9 million by 2023. During this pandemic of COVID 19 in India, 88% of internet users use social networking applications. The time spent on social networking applications across India increases from 3 hours and 13 minutes in Jan–Feb 2020 to 4 hours and 39 minutes in March–April 2020 (Statista 2020). During the first phase of lockdown, the usage of Voice Over Internet Protocol (VOIP) chat applications increases up to 5 hours and 17 minutes from 3 hours and 45 minutes pre-COVID-19 that was in Jan–Feb 2020 (Statista 2020).

A mental state characterized by the sense of fear or nervousness whenever being encountered with any problem is termed anxiety. During this time of the Pandemic of COVID 19, where the whole world is under lockdown, and we can say the earth stopped revolving, everything comes to a halt, People are under house arrest, and nobody is on the road. Offices are shut down, and in one way nation is breathing, but the economy lost its pulse. On the other hand, social media plays a significant role in being one of the primary sources of information about near surroundings and faraway places. Television and radio also covered a huge market, but social media has more of a connection with people, as, on social media, any news is trolled by the person who is already in your contacts. Further, another benefit of social media is getting information as and when required. Simultaneously, social media usage can attain social benefits, which leads to increased life satisfaction, and on, the usage of social media can also bring negative results through social overload (Zhan et al. 2016). The features of social media, ambient awareness and knowledge sharing are interrelated (Zhao et al. 2020). Facebook, Twitter, Instagram, WhatsApp, and LinkedIn are major social networking platforms used by a large population. Various posts uploaded over these platforms have their impact, and these may be in the form of pictures, thoughts, self-views etc. All form of posts has a varied level of interest creation over the web. Giving an emotional appeal to Facebook posts is one of the effective social media strategies for business and service marketers (Swani et al. 2013). The awareness regarding brand, word-of-mouth publicity, and finally purchase intentions have a positive impact through consumers' engagement with Facebook fan pages (Hutter et al. 2013). Various decisions were taken on the influence of what has been trolled over Facebook fan pages. The purchase decision process, which has multiple conscious or unconscious phases, is influenced by the various activities done on social media (Hutter et al. 2013). Social media can affect the rate of penetration of information to the public. The influence of social media is powerful in converting intentions into final buying. The spreading of real-time data during crises can be accelerated by the association of social media with end-users (Hornmoen and Maseide 2018).

Wuhan city of China turned out to be the epicentre of the spread of COVID-19 in December 2019 (Holshue et al. 2020). About 40 cases of pneumonia were reported, out of which few were working in the Huanan Seafood market. The World Health Organization (WHO) and the Chinese authorities worked together and gave the name Novel Corona Virus (2019-nCoV) to this new virus. On 11th January, a man, 61 years of age was reported as the first death due to this virus, and this person was also exposed to that seafood market. Within significantly less time, this virus spread across the world (WHO 2020). Outside China, the first death was reported on 2nd Feb 2020 in the Philippines. The name COVID-19 was given to this new virus disease by WHO on 11th February. After it was spread to 114 countries, WHO declared it a pandemic (WHO 2020). Due to the fast spread of this virus, COVID-19 affected a large population in a brief period. As of 7:12 pm, 27th May 2020, a total of 5,491,678 confirmed cases of COVID-19 globally were reported, which includes 3, 49,190 deaths reported by the WHO. In India total number of confirmed cases reported was 1, 51,767 which consisted of 4,337 deaths (WHO 2020).

During the technology era, the increasing use of social media brings myths along with fake news.

For a specific type of individual, these kinds of fake news were bothersome (Roy et al. 2020). The WHO, along with other governmental websites, works in clarifying this news along with the supply of authentic communication (WHO 2020). It is a standing instruction from the government not to supply any information without having an authenticity check. Globally every individual is affected by the spread of this virus differently (Roy et al. 2020). It has been found that the people who were isolated or kept under quarantine experience a significant level of stress like anxiety and post-traumatic stress symptoms (Brooks et al. 2020).

12.2 Literature Review

During the situations of disasters such as the Earthquake in Japan and Tsunami in 2011 and Typhoon Haiyan in the Philippines in 2013, social networking sites contributed significantly, and critical information regarding disasters was shared through social networking sites (Yi and Kuri 2016).

The time when any infectious disease spread in the community, social networking sites give a platform to share information with family and friends regarding the same. In the year 2009 during the spread of the H1N1 virus spread Twitter and Facebook played a key role in spreading information and experiences among the users (Chew and Eysenbach 2010, Signorini et al. 2011).

In the global crisis, social networking sites have become the main platform for risk communication. It not only speeds up information sharing but also gives the latest updates on a real-time basis (Ding and Zhang 2010).

Before the crisis, SNS was utilized by governments and organizations for proactive crisis management which includes sharing information regarding preparedness and various provisions regarding public health issues along with the feedback of the same. During the H1N1 spread in the United States in 2009, The Centres for Disease Control and Prevention (CDC) used Facebook and Twitter social networks to communicate about the spread (Ding and Zhang 2010).

This kind of approach makes organizations set up a positive reputation, through which the public at large can be made aware of the pandemic and their preparedness to counter those pandemics (Wan and Pfau 2004). Whenever a situation of pandemic occurs, generally people are willing to share their understanding and thoughts regarding the same with their friends and family members and even on some platforms to strangers too. In these situations, social networking sites play a significant role in engaging the expressive behaviours of individuals. During the spread of infectious disease, it was found that individual expresses their thoughts, views, and concerns by posting them on social networking sites. (Signorini et al. 2011, Vos and Buckner 2016). Largely we consider social networking sites as the platform to express these concerns, but it promotes online message reception also, apart from expressing people reading them too. But if we compare both behaviors, the behaviour to expression is active behaviour but the receptive behaviour is passive. During the pandemic, this receptive behaviour increases phenomenally (Weeks and Holbert 2013). It was found that during the time of H1N1 in 2009, the tweets increased from 8.8% to 40.5%, and over 2 million Twitter posts contained the keyword "H1N1" or "Swine flu" (Chew and Eysenbach 2010) and similar changed pattern regarding Swine flu-related tweets was also mentioned by McNeill, Harris, and Briggs (2016).

On the other hand, the darker side of this information sharing and receiving through social networking sites contributes to misinformation or adulteration of information is a matter of concern (Shin et al. 2018). Especially in the case of public health, this kind of information adulteration may lead to the creation of fear and suspicion among people, which can harm individual and societal well-being (Chen et al. 2015). As per a study on meta-analysis, during the Zika virus pandemic, it was found that over social networking sites the misinformation was more popular than the relevant information regarding the disease. Frequent sharing of information among users can generate misinformation (Sharma et al. 2017).

Rumor tweets or fake news instigate negative emotions in the users. Generally, in the normal case, the spread of rumors took some time to spread but in the case of any disaster, the spread of rumor regarding the same took very little time to spread (Miyabe et al. 2014). In a study regarding "Public Anxiety Toward Television Report on Airplane Accidents", it was concluded that any negative media report results in a rapid increase in public's anxiety and this air travel-related anxiety shoots up when the users also get exposed to the video reports related to the accident (Wang and Shu 2013). The major predictor of social media lassitude is the intensity to which a user uses social media. Comparison between social media and self-disclosure were major predictors for social media tiredness (Malik et al. 2020). Various studies have mentioned the negative psychological and behavioural outcomes of the feeling of social media tiredness, these are emotional exhaustion (Lim and Choi 2017), depression and anxiety (Pontes 2017, Dhir et al. 2018), sleep deprivation, concentration issues and relationship management problems (Salo 2017). Overloading information through social media has a strong association with social media lassitude (Whelan et al. 2020). Along with the information overload and social comparison the system's technical factors are also majorly related to social networking fatigue (Xiao et al. 2019).

In a study, it was estimated that more than 80% of the respondents were having a high frequency of exposure to social media. During COVID-19 the people who are frequently associated with social media is having a high possibility of developing anxiety and mental health problems (Geo et al. 2020). Health problems can be caused due to the misinformation overload on social media (Bontcheva et al. 2013, Roth and Brönnimann 2013). Basis elements of anxiety and stigma that generates information adulteration and rumors, majorly through social media were given by WHO (COVID W 2019). During the MERS outbreak in South Korea, it was found that social media exposure positively related to forming risk perceptions in the users (Choi et al. 2017).

This is the duty of practitioners and officials from public health to keep an eye on the diffusion of such misinformation or information adulteration through social networking sites among the public, especially the one who is less informed about the pandemic. For the same, they need to identify rumors and wrong information and provide accurate information through social media as well as traditional media. Further to that, they need to educate individuals regarding the prospective threats from the pandemic and the necessary safety measures one should take to secure themselves from the problem (Yoo and Choi 2019).

12.3 Procedure and Measure

Both the primary and secondary data were collected to frame the blueprint as well as construct the questionnaire. Secondary data was collected from various sources like research journals from reputed publication houses. More than 100 studies were consulted about the role of social media in causing stress, tension, anxiety etc. In these studies, as in social media, the study was conducted on Twitter, Facebook and WhatsApp in general. Few of the studies were conducted on pandemics or disasters. Out of these studies 32 studies were directly mentioned in this work. Other reports of MHA (Ministry of Home Affairs), reports on various data sources regarding the number of internet users, etc., were also consulted through Statista 2021.

The primary survey holds much weight in transforming the conceptual framework which was created through a strong literature review into notable findings about the given sample. The survey was done through an online mode; a Google form was created to add convenience for the researcher as well as respondents. The scale was divided into three sections, first section consists of the demographic features of the respondents (Table 1). The second section comprises the statements related to various roles of social media and the third section carries the statements related to the various feelings concerning the anxiety level of respondents. To check the validity and reliability of the self-administered scale, the following steps were taken:

1. The statements created in the scale were first reviewed by two experts in the domain area. This was an expert validity check, after which seven statements were deleted based on lack of clarity and ambiguity in nature.

Table 1: Demographic profile of Respondents.

Sr. No.	Demographic characteristics of respondents	Participants (N = 327)
1. Gender (Respondents of age 18 to 25 years)		
	Male	163
	Female	164
2. Occupation		
	Student	227
	Government Job	46
	Private Job	54
3. Marital Status		
	Married	104
	Unmarried	223

Source: Primary Survey

2. The pilot study was conducted on the remaining statements after face validity with 50 respondents. The reliability value, i.e., Cronbach α was found 0.78, as per Cuieford (1965), a Cronbach α coefficient greater than 0.7 is considered as highly reliable.

The final data was collected through the web survey, the study was conducted on the respondents of Delhi - NCR. Five-point Scale was used (1= Strongly Disagree to 5 = Strongly Agree). The scale was distributed to 700 respondents and out of which the final count received was 327, which was further used for the analysis. The snowball sampling technique was used.

12.4 Analysis

To address the research problems mentioned above, exploratory factor analysis (EFA) and simple Regression techniques were applied for checking the usage of social media and its effect on anxiety among youth.

Since the study is confined to young adults covering the age group of 18 to 25 years. The demographic profile of these respondents reveals that both the genders accounted equal and in terms of occupation the sample has a spread as students, government, and private jobs and 69.4% of the respondents are students. As per marital status, 68.2% of the respondents are unmarried.

For analysis, exploratory factor analysis is a technique used for exploring the factor by reducing the number of statements. Reduction and association of statements in a particular factor derive the conclusive outcome from the statements. After an extensive literature review, it was found that the factors of the role of social media and anxiety need to be explored for further understanding.

Table 2 presents the statistical value for the justification of the said technique, KMO value is 0.826, which justifies the feasibility of the application of factor analysis on the given data. Results of Bartlett the test of sphericity allow for rejecting the null hypothesis according to which the original correlation matrix would be unitary (approx. Chi-square 5860.413 df = 378, Sig. = 0.000).

Thus, both the results of KMO and Bartlett's Test found that data is appropriate and adequate for applying factor analysis. Other assumptions regarding the application of factor analysis are also fulfilled. Further, the total variance explained with six factors collectively comes out to 66.33%.

Factor Loading gives the clear inclusion of items in factors. Table 3 shows the factor loading of each variable after rotation through the Varimax procedure and the rotation is converged in eight iterations. No theory clearly states the minimum level of factor loading to be considered

Table 2: KMO and Bartlett's Test.

KMO and Bartlett's Test		
Kaiser-Meyer-Olkin Measure of Sampling Adequacy.		0.826
Bartlett's Test of Sphericity	Approx. Chi-Square	5860.413
	Df	378
	Sig.	0.000

Source: Primary Analysis

but factor loadings greater than 0.5 should be considered adequate for further analysis (Nunnally (1978), Malhotra and Dash (2016)), and therefore, factor loadings having a value below 0.5 has been neglected in the present study.

As per Table 4, the combination of various statements under a particular factor as per their loadings is named Information Source (social media as an Information source), Professional Connect (social media as Professional Connect), Entertainment (social media as a source of Entertainment), Physical Anxiety, Social Anxiety, Cognitive Anxiety.

The forthcoming section explains the effect of various social media roles on anxiety building. In Table 5, all the three regression models were explained in which Physical Anxiety, Social Anxiety and Cognitive Anxiety are considered as dependent variables and all three roles of social media, i.e., as a source of Information, as a medium of Professional Connect and as a source of Entertainment as an independent variable. The model summary is followed by the three equations explaining the weights of all the independent variables in all three equations.

1. Physical Anxiety = 0.186 + 0.205 (Source of Information) + 0.366 (Source of Entertainment)

2. Social Anxiety = 0.212 + 0.212 (Source of Information) – 0.161 (Source of Professional Connect) + 0.463 (Source of Entertainment)

3. Cognitive Anxiety = 0.206 + 0.313 (Source of Information) + 0.232 (Source of Entertainment)

As per equation number 1, there are two major contributors to the building of physical anxiety (a) social media as a source of information and (b) social media as a source of entertainment. Both of these came out as significant contributors (p value is less than 0.05) whereas the third role which is social media as a professional connection is observed as Insignificant. It depicts that young adults are spending lots of time on social media and show a piece of information and entertainment-seeking behaviour.

Table 3: Rotated Component Matrix: Social-Media & Anxiety.

Rotated Component Matrix						
	Component					
	1	2	3	4	5	6
Social media spreads information about the effects, symptoms and causes of COVID 19 effectively.	0.781					
I usually seek information about the number of COVID cases through social media.	0.728					
During COVID 19, getting the information regarding the wellbeing of near ones is effective through social media only.	0.714					
For me during this pandemic, receiving and spreading information through social media is comparatively easy and effective.	0.709					
During COVID 19, Scial Media is the only source for information transfer	0.636					
I often feel body ache in the back during this time						
For me social media like Linkedin is a source of serious professional connect.		0.822				
During pandemic professional linkups grows through social media only.		0.78				
For me social media is for socializing with professional links and it has proved during lockdown period.		0.767				
Job losses during COVID 19 was countered by better reach to various job providers through social media.		0.766				
During lockdown, when physical connect was restricted, social media plays a big role in professional linkups.		0.718				
I often feel weak and tired during this time			0.753			
I often feel an increase in the heartbeat during this time			0.705			
I often feel a sense of nausea/vomiting during this time			0.705			
I am unable to relax myself			0.648			

Table 3 contd. ...

...Table 3 contd.

Rotated Component Matrix						
	Component					
	1	2	3	4	5	6
I often feel a high body temperature during this time			0.637			
Virtual connect is another form of social connect.						
For me shared content over social media is the source of entertainment during lockdown.				0.867		
Snapchat, Twitter, Instagram proved to be an effective source of entertainment during COVID 19 Pandemic.				0.849		
During lockdown the time spent for entertainment increased enormously over social media.				0.778		
I feel tensed due to the current situation in society					0.789	
I feel I get affected due to spread of COVID 19					0.741	
I feel worried to step out of home					0.656	
During COVID 19, getting the information regarding the wellbeing of near ones is effective through social media only.						
I am facing a lack of pleasure in my hobbies						0.707
It disturbs my sleeping pattern						0.67
I often suffer from indigestion during this time						0.668
My frequency of information sharing through social media increased during COVID 19						

Extraction Method: Principal Component Analysis.
Rotation Method: Varimax with Kaiser Normalization[a]
a. Rotation converged in 8 iterations.

Falk and Miller (1992) recommended that any value of R square equal to or more than 0.10 is considered adequate for explaining the dependent variable variance. As per Cohen (1998) the R square values are categorized as:

0.26—Substantial, 0.13—Moderate & 0.02—Weak

Table 4: Nomenclature of Factors.

Sr. No	Statements	Nomenclature
1	Social media spreads information about the effects, symptoms and causes of COVID 19 effectively	Information Source (social media as Information source)
2	I usually seek information about the number of COVID cases through social media	
3	During COVID 19, getting the information regarding the wellbeing of near ones is effective through social media only	
4	For me during this pandemic, receiving and spreading information through social media is comparatively easy and effective	
5	During COVID 19, Scial Media is the only source for information transfer	
6	For me social media like Linkedin is a source of serious professional connect	Professional Connect (social media as Professional Connect)
7	During pandemic professional linkups grows through social media only	
8	For me social media is for socializing with professional links and it has proved during lockdown period	
9	Job losses during COVID 19 was countered by better reach to various job providers through social media	
10	During lockdown, when physical connect was restricted, social media plays a big role in professional linkups	
11	I often feel weak and tired during this time	Physical Anxiety
12	I often feel an increase in the heartbeat during this time	
13	I often feel a sense of nausea/vomiting during this time	
14	I am unable to relax myself	
15	I often feel a high body temperature during this time	
16	For me shared content over social media is the source of entertainment during lockdown	Entertainment (social media as source of Entertainment)
17	Snapchat, Twitter, Instagram proved to be an effective source of entertainment during COVID 19 Pandemic	
18	During lockdown the time spent for entertainment increased enormously over social media	
19	I feel tensed due to the current situation in society	Social Anxiety
20	I feel I get affected due to spread of COVID 19	
21	I feel worried to step out of home	
22	I am facing a lack of pleasure in my hobbies	Cognitive Anxiety
23	It disturbs my sleeping pattern	
24	I often suffer from indigestion during this time	

Table 5: Summary of Regressions.

Sr. No.	Model Summary (Adj R^2 and Durbin Watson)	ANOVA (F Value and Sig.)	Variables	
			Dependent Variable	Independent Variables (Standardized Coefficients β; Sig.)
1	23.9%; 2.187	35.04; .000	Physical Anxiety	• Social Media as Source of Information (0.205; 0.003) • Social Media as a Professional Connect (0.092; 0.174) • Social Media as Source of Entertainment (0.366; 0.000)
2	24.9%; 2.287	37.029; .000	Social Anxiety	• Social Media as Source of Information (0.212; 0.002) • Social Media as a Professional Connect (-0.161; 0.017) • Social Media as Source of Entertainment (0.463; 0.000)
3	19%; 1.862	26.543; .000	Cognitive Anxiety	• Social Media as Source of Information (0.313; 0.000) • Social Media as a Professional Connect (0.038; 0.585) • Social Media as Source of Entertainment (0.232; 0.000)

In equation 1, both these factors collectively explain 23.9% of physical anxiety which is very much close to substantial.

Similarly in equation 2, Source of Information, Source of Professional Connect and source of Entertainment explain 37% of social anxiety which is substantial. In the current equation, social anxiety is found to be negatively correlated with Source of Professional Connect, which means a 1% increase in Physical and Social Connect may lead to a 0.161% decrease in the Social Anxiety level. Further, the other two variables are positively correlated.

Equation 3, the contributors to cognitive anxiety are Source of Information and Source of Entertainment and both these independent variables collectively could explain 26.54% of Cognitive anxiety which again comes out substantial.

Therefore, after knowing the impact of various social media roles on all three different kinds of anxieties, the collective impression of these independent variables over total anxiety needs to be studied.

Table 6 exhibits values of R square and adjusted R square. The results indicate that all three roles of social media collectively can explain 35 % (adjusted R Square) of variance in total anxiety (dependent variable). Further, it implied that 65% of Anxiety created during a pandemic could be explained by something other than the social media concept. As per

Table 6: Effect of Social media usage on Aggregate Anxiety.

Model Summary[b]					
Model	R	R Square	Adjusted R Square	Std. Error of the Estimate	Durbin-Watson
1	.597[a]	0.356	0.35	0.46724	2.069

a. Predictors: (Constant), Social Media as Source of Entertainment, Social Media as a Professional Connect , Social Media as Source of Information
b. Dependent Variable: Total Anxiety

Table 7: Coefficients[a.]

	Model B	Unstandardized Coefficients		Standardized Coefficients	t	Sig.
		Std. Error	beta			
1	(Constant)	1.877	0.142		13.232	0.000
	Social Media as Source of Information	0.324	0.032	0.466	10.257	0.000
	Social Media as a Professional Connect	−0.01	0.039	−0.016	−0.255	0.799
	Social Media as Source of Entertainment	0.198	0.04	0.309	4.908	0.000

a. Dependent Variable: Total Anxiety

Durbin and Watson (1950) if the value of Durbin-Watson comes out less than 2 then there will be positive autocorrelation and in the case of more than 2 the negative autocorrelation exists, in the present case, the value of Durbin-Watson comes out 2.069 which is nearly equal to 2 and shows zero autocorrelation.

It can be witnessed from above Table 7 that the role of social media as a source of information has shown the maximum effect on Anxiety building with a standardized beta value (β = .466) followed by social media as a source of entertainment (β = 0.309) and least influence was due to the third predictor of Anxiety that is social media as a Professional Connect (β = −0.016). This predictor has a negative influence, although the third predictor comes out insignificant, the p value is greater than 0.05. So, the impact of different roles of social media on anxiety may be explained as under:

Total Anxiety = 1.877 + 0.466 (social media as Source of Information) + 0.309 (social media as Source of Entertainment)

12.5 Conclusion

During the Pandemic COVID-19, social media use has grown exponentially. Along with the satisfaction either through information seeking, professional connection or entertainment, social media also brings unrest among the users during the pandemic, termed anxiety. Under this study, young adults were targeted as respondents, and the effect of different roles of social media is assessed over anxiety building. During the primary investigation, it was found that social media influences anxiety building. Through factor analysis, it is explored that anxiety has three major components: physical anxiety, social anxiety, and cognitive anxiety. Social media's roles can be further bifurcated as a source of information, a source of professional connection and a source of entertainment.

Further, it is found that Physical Anxiety is caused by social media usage as a source of entertainment. Using social media as a source of entertainment during a pandemic when other outdoor sports or other entertainment options are restricted calls for a sedentary lifestyle which finally causes physical unrest, and this physical unrest causes physical anxiety. Similarly, this will also affect the social interaction between the people and restrict them to themselves and create Social Anxiety. Social media usage as a source of professional connection will inversely affect social anxiety. The increase in the use of social media for Professional connection will lead to a reduction in social anxiety. The information spread over social media affects building cognitive anxiety.

When we analyze it collectively, the effect of various roles of social media over anxiety building in total, the role as a source of information comes out as dominating factor leads as a cause of anxiety building. Information overload may be the reason for the creation of anxiety. These results support the findings of Bontcheva et al. (2013) and Roth and Brönnimann (2013), that health problems can be caused due to the misinformation overload on social media. Further, as per Whelan et al. (2020), information overload through social media has a strong association with social media lassitude, a state of mental and physical weakness.

12.6 Limitations of Tthe Study

This study was conducted on selected respondents of one region, where the population, after being diverse, shares some common regional traits. The young adults of another area may have a different appetite for anxiety-bearing. Therefore, it would not be easy to generalize the results for all young adults country-wide. The availability of finance and time also acts as a limiting factor for the study.

12.7 Future Scope for the Study

Further research can be conducted by using the exclusive scale of Anxiety. As Anxiety is a Psychological Phenomenon, the study demands a controlled environment. The study may also be extended to the specific working domains, and pre-and post-pandemic results can also be compared.

References

Bontcheva, K., G. Gorrell, and B. Wessels. 2013. Social media and information overload: Survey results. *arXiv preprint arXiv:1306.0813*.

Brooks, S. K., R. K. Webster, L. E. Smith, L. Woodland, S. Wessely, N. Greenberg, and G. J. Rubin. 2020. The psychological impact of quarantine and how to reduce it: rapid review of the evidence. *The Lancet*.

Chen, X., S. C. J. Sin, Y. L. Theng, and C. S. Lee. 2015. Why students share misinformation on social media: Motivation, gender, and study-level differences. *The Journal of Academic Librarianship* 41(5): 583–592.

Chew, C., and G. Eysenbach. 2010. Pandemics in the age of Twitter: content analysis of Tweets during the 2009 H1N1 outbreak. *PloS one* 5(11).

Choi, D. H., W. Yoo, G. Y. Noh, and K. Park. 2017. The impact of social media on risk perceptions during the MERS outbreak in South Korea. *Computers in Human Behavior* 72: 422–431.

COVID, W. 2019. PHEIC Global research and innovation forum: towards a research roadmap.

Dhir, A., Y. Yossatorn, P. Kaur, and S. Chen. 2018. Online social media fatigue and psychological wellbeing—A study of compulsive use, fear of missing out, fatigue, anxiety and depression. *International Journal of Information Management* 40: 141–152.

Ding, H., and J. Zhang. 2010. Social media and participatory risk communication during the H1N1 flu epidemic: A comparative study of the United States and China. *China Media Research* 6(4): 80–91.

Gao, J., P. Zheng, Y. Jia, H. Chen, Y. Mao, S. Chen, Y. Wang, H. Fu, and J. Dai. 2020. Mental health problems and social media exposure during COVID-19 outbreak. *Plos One* 15(4): e0231924.

Holshue, M. L., C. DeBolt, S. Lindquist, K. H. Lofy, J. Wiesman, H. Bruce, and G. Diaz. 2020. First case of 2019 novel coronavirus in the United States. *New England Journal of Medicine*, 929–936.

Hornmoen, H., and P. H. Måseide. 2018. Social Media in Management of the Terror Crisis in Norway: Experiences and Lessons Learned.

https://www.statista.com/topics/1164/social-networks/#topicHeader__wrapper. Published by S. Dixon, Jun 21, 2022.

Hutter, K., J. Hautz, S. Dennhardt, and J. Füller. 2013. The impact of user interactions in social media on brand awareness and purchase intention: the case of MINI on Facebook. *Journal of Product & Brand Management*.

Lim, M. S., and S. B. Choi. 2017. Stress caused by social media network applications and user responses. *Multimedia Tools and Applications* 76(17): 17685–17698.

Malik, A., A. Dhir, P. Kaur, and A. Johri. 2020. Correlates of social media fatigue and academic performance decrement. *Information Technology & People*.

McNeill, A., P. R. Harris, and P. Briggs. 2016. Twitter influence on UK vaccination and antiviral uptake during the 2009 H1N1 pandemic. *Frontiers in Public Health* 4: 26.

Miyabe, M., A. Nadamoto, and E. Aramaki. 2014. How do rumors spread during a crisis? Analysis of rumor expansion and disaffirmation on Twitter after 3.11 in Japan. *International Journal of Web Information Systems* 10(4): 394–412.

Pontes, H. M. 2017. Investigating the differential effects of social networking site addiction and Internet gaming disorder on psychological health. *Journal of Behavioral Addictions* 6(4): 601–610.

Roth, F., and G. Brönnimann. 2013. Focal report 8: risk analysis using the internet for public risk communication. Risk and Resilience Research Group. *Center for Security Studies (CSS), ETH Zürich.*

Roy, D., S. Tripathy, S. K. Kar, N. Sharma, S. K. Verma, and V. Kaushal. 2020. Study of knowledge, attitude, anxiety & perceived mental healthcare need in Indian population during COVID-19 pandemic. *Asian Journal of Psychiatry*, 102083.

Salo, J. 2017. Social media research in the industrial marketing field: Review of literature and future research directions. *Industrial Marketing Management* 66: 115–129.

Sharma, M., K. Yadav, N. Yadav, and K. C. Ferdinand. 2017. Zika virus pandemic—analysis of Facebook as a social media health information platform. *American Journal of Infection Control* 45(3): 301–302.

Shin, J., L. Jian, K. Driscoll, and F. Bar. 2018. The diffusion of misinformation on social media: Temporal pattern, message, and source. *Computers in Human Behavior* 83: 278–287.

Signorini, A., A. M. Segre, and P. M. Polgreen. 2011. The use of Twitter to track levels of disease activity and public concern in the US during the influenza A H1N1 pandemic. *PloS one* 6(5).

Swani, K., G. Milne, and B. P. Brown. 2013. Spreading the word through likes on Facebook. *Journal of Research in Interactive Marketing.*

Vos, S. C., and M. M. Buckner. 2016. Social media messages in an emerging health crisis: Tweeting bird flu. *Journal of Health Communication* 21(3): 301–308.

Wan, H. H., and M. Pfau. 2004. The relative effectiveness of inoculation, bolstering, and combined approaches in crisis communication. *Journal of Public Relations Research* 16(3): 301–328.

Wang, W., and S. Cole. 2013. Public Anxiety Toward Television Report on Airplane Accidents', Advances in Hospitality and Leisure (Advances in Hospitality and Leisure, Volume 9).

Weeks, B. E., and R. L. Holbert. 2013. Predicting dissemination of news content in social media: A focus on reception, friending, and partisanship. *Journalism & Mass Communication Quarterly* 90(2): 212–232.

Whelan, E., A. N. Islam, and S. Brooks. 2020. Is boredom proneness related to social media overload and fatigue? A stress–strain–outcome approach. *Internet Research.*

World Health Organization. 2020. WHO Director-General's Opening Remarks at the Media Briefing on COVID-19-11 March 2020.

Xiao, L., J. Mou, and L. Huang. 2019. Exploring the antecedents of social network service fatigue: a socio-technical perspective. *Industrial Management & Data Systems.*

Yi, C. J., and M. Kuri. 2016. The prospect of online communication in the event of a disaster. *Journal of Risk Research* 19(7): 951–963.

Yoo, W., and D. H. Choi. 2019. Predictors of expressing and receiving information on social networking sites during MERS-CoV outbreak in South Korea. *Journal of Risk Research*, 1–16.

Zhan, L., Y. Sun, N. Wang, and X. Zhang. 2016. Understanding the influence of social media on people's life satisfaction through two competing explanatory mechanisms. *Aslib Journal of Information Management.*

Zhao, Y., X. Zhang, J. Wang, K. Zhang, and P. O. de Pablos. 2020. How do features of social media influence knowledge sharing? An ambient awareness perspective. *Journal of Knowledge Management.*

CHAPTER 13

An Exploratory Study of Understanding Consumer Buying Behaviour Towards Green Cosmetics Products in the Indian Market

Md. Sohail,[1] Richa Srivastava[2] and Srikant Gupta[1],*

Contents

13.1 Introduction

13.2 Literature Review

13.3 Research Methodology

13.4 Result Analysis

13.5 Conclusion

13.1 Introduction

As the COVID-19 pandemic has made us hit rock bottom, it has become extremely important for us to analyse the impact of our actions on the environment around us. And, while doing so, it becomes extremely essential to identify the factors that have a huge influence on deteriorating the environmental setting we live in. When we observed the various industries, we came across one industry that might not be seen as directly contributing to worsening the ecosystem, but contributes to it in several

[1] Jaipuria Institute of Management, Jaipur.
[2] Jaipuria Institute of Management, Lucknow.
* Corresponding author: srikant.gupta@jaipuria.ac.in

ways. It is, in fact, the cosmetics industry, which we rarely see as negatively affecting the environment (Ghazali et al. 2017).

However, the growing attraction of the masses to this industry has worsened the situation. One might wonder how the seemingly harmless products might lead to damage if any. Yet, research shows that the toxic chemicals present in cosmetics have led to damage to the ocean ecosystem, and, further, to the deaths of aquatic species. Livestock that comes into contact with the toxins that end up in the soil might even suffer from various abnormalities, be they reproductive, genetic, or development, or might even have cancer. Not only do cosmetics deplete natural resources, but the packaging itself takes hundreds of years to decompose, thereby, further aggravating the situation. One of the major components of makeup is palm oil, which has been contributing to rapid deforestation and climate change. Many other components, such as phthalates, Triclosan, Butylated Hydroxyanisole, Diethanolamine, and so on, also harm the environment in one way or the other. (Tewari et al. 2022).

Therefore, it has become extremely essential to address these issues. A major change can be brought about by consumers. If they shift to green cosmetics, then these issues can be addressed one by one. Even though there has been a gradual shift, the pace is slow, and it might take some time for the harmful effects to be addressed. Cosmetic companies have an advantage too by going green, such as a good brand reputation, new target segmentation, an opportunity to enter foreign markets, and a chance to enter the premium segment of the market. Increasing consciousness and concern for health and the atmosphere are increasingly changing consumer behaviour toward environmentally sustainable goods. According to the advertisers' eco-friendly goods have a minimal negative impact on the environment.

13.2 Literature Review

Rawat and Garga (2012) explored the potential of green marketing in the cosmetics industry to meet the needs of everyone from newborns to young people. A sample of 200 working women in Pune yielded quite optimistic results. From environmentally friendly cosmetics to chemicals, relatively high costs are seen as a threat, but changes in India's demographic and economic factors have also proved this. Bruno (2015) addressed that green marketing is an overall marketing concept in which the production, sales, consumption, and disposal of products are environmentally friendly. Words such as "recyclable", "naturally derived", "good for ozone", "paraben-free" and "phosphate-free" indicate environmentally friendly marketing. According to Born et al. (1989), cosmetics have been used by humans since the dawn of time. As far back as 10,000 BC, there was a time when men and women painted their noses, rouged their lips and cheeks, and used Heena to paint their nails with a lot of kohl on the brow's, kohl was a dark-coloured powder formed from powdered antimony or some mixture of these materials. Sharma et al. 2021, discussed the effect and awareness level of green cosmetics products among customers buying green items in the Delhi NCR region. This study uncovers that characteristic concern; self-significance impacts green buying while personal development has an adverse effect. Furthermore, fragment factors have a basic sway on green getting, one intriguing circumstance here for sexual direction is that males are shown to be more ideal for eco-obliging things than females, most various assessments ensure regardless for instance depicting females all the more useful for eco-obliging things. Green stamping and eco-marks are a segment of the critical factors influencing customers' attitudes towards green things while green publicizing and green packaging accepted an immaterial part (Amberg and Fogarassy (2019) found that environmental concern and health consciousness have a major impact on green buying for cosmetics, and indicate that women have shown more openness than men in buying green cosmetics. Around 70% of the participants showed their willingness to buy green cosmetics in the future. Askadilla and Krisjanti (2017) observed that attitude towards green cosmetics and how they perceived the behavioural control have a significant influence on whether to buy green cosmetics products. Among all these factors, attitudes showed the most significance. Also, the study tells the marketers to not just focus on the internal benefits and development of the product but also to work on creating environmental awareness among the consumers to further develop the market for green cosmetics. Testa et al. (2022), Consumers' increased environmental knowledge, as well as the complexity of environmental concerns, need more research into how they deal with such logic. In the shift towards a more circular economy, a better understanding of the complexities of consumer decision-making processes and how contradictory logic might cause tensions in purchasing decisions is critical. It is the goal of this study to investigate how important factors like recycled materials, brand loyalty, and plastic concern might be used to predict how customers will act Sadiq et al. (2021). This study explores consumer resistance to buying eco-friendly cosmetics and explores the moderating roles of environmental and health concerns. 350 consumers were surveyed online, and the data showed that all the hurdles are substantial deterrents to using eco-friendly cosmetics. The results suggest that tradition and image barriers have the greatest impact on consumers' purchase intentions. The negative influence of value and image barriers diminishes customers' intention to buy eco-friendly things, whereas health concerns reduce the influence of tradition and risk barriers Sreen et al. (2021). In a study, it's found that consumers are sceptical of brands' claims of naturalness. It's critical to understand the causes of loyalty and the underlying brand love for such items. The study created a conceptual model based on the behavioural reasoning theory to investigate the antecedents of brand love in response to this demand. Researchers used a cross-sectional survey to gather data, implying that health consciousness is linked to motives for purchasing natural products and attitudes, both of which are linked to each other. At the same time, the reasons for consuming natural products and one's attitude toward them are both favourably connected with brand loyalty to natural products. On the other hand, the reasons why people don't use natural products are all about attitude and have nothing to do with health or brand loyalty.

13.3 Research Methodology

13.3.1 Research Problem

Through this research, we aim to address the problem and to learn about the factors that influence the buying pattern of people who are willing to make green purchases or are already active green purchasers so that the same learnings can be applied to the market of green cosmetics. So, the research problem would be "*An exploratory study of understanding consumer buying behaviour towards green cosmetics products in the Indian Market*". Also, the research questions underlying the research problem can be described as follows:

- Which demographic and psychographic factors influence the preference for green cosmetic products and to what extent do these factors influence?
- Is there any relationship that exists between green consumerism and high social status?
- Does green marketing influence the purchase preferences of consumers for green cosmetic products?

13.3.2 Significance of the Research

Generally, the studies related to this topic are based on cosmetics, as the research papers on the topic of green cosmetics are significantly fewer in number, so with the help of this research paper, the motive will be to contribute to the development of this field. Recently, for the last few decades, male cosmetics have been seen as important by the people who sell them. This marketing has been mostly done with the influence of social media, and celebrity endorsements, which play a greater role in spreading its concept. The literature is important when you want to look at the factors that make people buy green cosmetics based on their female attitudes. The main purpose of this chapter is to study both males and females, and because of these kinds of differences in the attitudes that exist between the genders, the study of both of them creates a proper strategy for males' green cosmetic behaviour as well, which is for a new growth avenue.

After reviewing the existing literature, which is based on this topic, we see that the research studies are less when compared with western world countries than India for this topic. Besides this, most of the research studies have focused on women's cosmetics only, whereas this study tries to study the buying behaviour of males and females. In addition to this, the knowledge of the author and a smaller amount of research done in India suggests that this research will help many beneficiaries in this process, as in this attempt to bring out the most beneficial information related to consumers and many other factors that will help influence a green cosmetic purchase.

The increase in the demand for cosmetics and green products creates a need for more successful studies that can be carried out in the future. The outcome of these studies might help societies, provided that these cosmetics have begun to be used in the daily lives of people. This also serves as a future reference for many researchers who want to investigate this subject in the future.

13.3.3 Management Decision Problems

- A lot of people are using cosmetics at present, but most of them are using harmful cosmetics, both for humans and the environment. So, here, we are trying to analyse the awareness of consumers towards green cosmetics, which are eco-friendly and also good for their health. Before proceeding further with the research, we have decided upon the following objectives, which will drive the research process. Awareness of Green cosmetics: Since we aim to analyse the purchase behaviour of the customers towards green cosmetics, the basic aim in mind is to analyse the awareness regarding these products in the minds of the customers.

- Impact of its environmental benefits: Green cosmetics are eco-friendly products, while regular cosmetics products harm the environment as well as humans. We, thereby, want to analyse whether the environmental benefits are considered by the consumers while making the purchase decision.

- Consideration of the life-saving benefits: Some cosmetics products are made from animals, such as tallow, gelatine, lanolin, oestrogen, and many more, which results in the death of millions of animals every year. This should aim to understand whether knowing this fact will have an impact on the minds of the consumers of cosmetics.

- Analysis of the side effects: Side effects are very common when using cosmetics, but these green cosmetics are made up of natural products. So, they have very few or no side effects. We, thus, want to know whether this would lead more consumers to shift to green cosmetics.
- Influence of the cost constraint: Many people think that green products are very expensive, but that is not true. Rather than this, it fits into the budget of the common people as well. Thus, researchers wish to analyse whether prices act as a major influencer in impacting the purchase decisions of the consumer.

13.3.4 Variables to the Objectives

I) Analysing demographic factors that affect consumer buying of green cosmetic products in the Indian region.
   Variables to study
   - V1: Gender;
   - V2: Age;
   - V3: Education level;
   - V4: Working status;
   - V5: Family income level.

II) Analyzing psychographic factors that affect consumer buying of green cosmetic products in the Lucknow region.
   Variables to study
   - V1: Sustainability;
   - V2: Safety and Quality;
   - V3: Brand connection;
   - V4: Pricing;
   - V5: High product involvement.

III) To analyse whether green consumerism leads to the sign of high social status.
   Variables to study:
   - V1: Buying green cosmetic products helps me express my personality;
   - V2: I am willing to sacrifice luxury and performance by buying green cosmetic products;
   - V3: Buying green cosmetic products will make me feel like a responsible citizen;

- **V4**: Buying green cosmetic products will make me look knowledgeable;
- **V5**: Buying green cosmetic products will make me look wealthy;
- **V6**: Buying green cosmetic products will make me look ethical;
- **V7**: You can tell a lot about a person from whether he/she buys green cosmetic products.

IV) To analyse whether green marketing has any impact on the consumer buying of Green Cosmetic Products.
Variables to study

- **V1**: Stores with a fresh and vibrant boutique look (sustainable furniture);
- **V2**: Stores with LED lighting and non-toxic paints;
- **V3**: Recycle packaging;
- **V4**: Environment campaigns;
- **V5**: Community building and giving back to society

13.3.5 Research Methods and Design

The sampling techniques that are used in this research study are judgmental sampling and snowball sampling. The attempts that were made to gather the respondents were limited to the users of cosmetics or those people who had prior knowledge of the cosmetics. All the research is based on green cosmetics, and in the first place, it needs to be made sure that the respondents are already users, and if they are familiar with any cosmetics, only then can they can give any views on green cosmetics. If the respondent is not familiar with the cosmetics, then the questions are based on the perception of non-existent green cosmetics. So this was the reason behind the usage of the judgmental sampling techniques, as they were used to create the characteristics of "use of familiarity with cosmetics" amongst the population.

For data collection, structured questionnaires were used and gathered the data online and the statistical analysis, which was done through ANOVA, descriptive statistics, T-Test, and multiple regression, was used for the findings. The first stage was used for simple statistical studies in which averages, percentages, etc., were used. After the descriptive research, analytical research was done by testing the hypothesis and interpreting the relationship. The use of advanced statistical techniques such as ANOVA and F-test coefficient of determination was used, and after doing analytical research and explanatory research, which is used in explaining the derived interrelationship in depth.

13.3.6 Data Collection and Segregation

Distinct exploration, Analytical examination, and Explanatory exploration are considered to be the most suitable exploration plan. A graphic exploration configuration will help in zeroing in on the specific part of the issue. It likewise incorporates what is the issue here, various ways through which information can be accumulated, handling of information, factual investigation of the information, and finally discoveries. Essential information was assembled through surveys and measurable investigations such as ANOVA, factor examination, and different relapse investigation was utilized for discoveries. At an early stage, basic factual investigations like midpoints, rates, and so on were utilized. After research, the scientific examination was finished by testing speculations and deciphering connections. Here cutting-edge measurable procedures, for example, ANOVA, F-test, and coefficient of assurance were utilized. After scientific examination, Explanatory exploration is utilized for clarifying the determined interrelationships from top to bottom.

The survey follows the case of going from general requests to being expressed. There are a total of 27 requests divided into 3 regions and 8 sub-fragments. The important subsection has requests concerning which restorative class a respondent buys the most, a repeat of its purchase, General parts, and Environmental factors for the purchase decision of beautifiers. The resulting subsection gets some data about do they comprehend what green thing is, expectations of purchasing green things, any past practical thing purchase, and their view about purchasing green things for a sensible future. The third subsection consolidates psychographic information for instance sees on security, suitability, and checking for Green improving specialists. The fourth subsection was related to the thing commitment of the purchasers, and the fifth included requests that were a feature of a thing, information, and openness of things.

13.4 Result Analysis

Demographics of the sample

This section presents the demographic information of 69 respondents that were surveyed through an online questionnaire for conducting this research study. The profile covers gender, age, education level, work status, and household monthly income. Based on the data, out of 69 respondents, 37 respondents (54%) are female and 32 respondents (46%) are male. The 45 respondents (65.2%) belong to the age group of 18–24, 13 respondents (18.8%) belong to the age group of 25–35, 6 respondents (8.7%) and 5 respondents (7.2%) belong to the age group of 45–54. Based on educational background, 16 respondents (23.2%) are Bachelor's degree graduates,

50 respondents (72.5%) are Master's degree graduates, 2 respondents (2.9%) are high school graduates and 1 respondent (1.4%) has doctoral degrees or more. A good number of respondents 51 (73.9%) are students, and 18 respondents (26.1%) are employed.

Influence of Demographic Profile on Buying of Green Cosmetic Products

Gender: Gender is a factor that might affect consumer buying of green cosmetic products. ANOVA is used to know if gender has any significant impact on the choice of green cosmetic products. The respondents are classified into two categories; (a) Male and (b) Female, respectively as 1 and 2 for SPSS analysis. The significant level (p-value) of 0.752 is more than the α value of 0.05. Since the p value is more than the α value, the null hypothesis is accepted. Hence, gender does not significantly impact the consumer buying of Green Cosmetic Products.

Education level: Education level is a factor that might affect consumer buying of green cosmetic products and is a dependent variable. ANOVA is used to know if education level has any significant impact on the choice of green cosmetic products. The respondents are classified into five categories; (a) High school passed students (b) Bachelor's degree holders (c) Master's degree holders (d) Doctoral students who graduated, respectively as 1, 2, 3 and 4 for SPSS analysis. The significant level (p value) of 0.533 is more than the α value of 0.05. Since the p value is more than the α value, the null hypothesis is accepted. Hence, education level does not significantly impact the consumer buying of green cosmetic products.

Work status: Work status is a factor that might affect consumer buying of green cosmetic products and is a dependent variable. ANOVA is used to know if work status has any significant impact on the choice of green cosmetic products. The respondents are classified into four categories; (a) Student (b) Employed, respectively as 1 and 2 for SPSS analysis. The significant level (p value) of 0.017 is less than the α value of 0.05. Since the p value is less than the α value, the null hypothesis is rejected. Hence, work status significantly impacts the consumer buying of green cosmetic products.

Household income: Household income (monthly) is a factor and consumer buying of green cosmetic products is a dependent variable. ANOVA is used to know if household income has any significant impact on the choice of green cosmetic products. The respondents are classified into seven categories; (a) No income, (b) Less than 25,000, (c) 25,000–49,999, (d) 50,000–74,999, (e) 75,000–99,999, (f) 1,00,000–150,000, (g) More than 150,000, respectively as 1, 2, 3, 4, 5, 6, and 7 for SPSS analysis. The significant level (p value) of 0.083 is less than the α value of 0.05. Since the p value is less than the α value, the null hypothesis is rejected. Hence, monthly household income significantly impacts the consumer buying of green cosmetic products.

Respondents' Knowledge, Buying Behaviour, Attitude related to buying Green Products

General knowledge of Green Products (including Cosmetics)

The result shows that 69 respondents, i.e., 50.7% of the respondents have knowledge about green cosmetic products, and as there is a considerably half number of respondents who are not aware of green cosmetic products, we will get relevant results for the research through the responses formed. People who are aware can give better insights through their experiences and hence better results than the people who are unaware in the first place.

Have you ever bought any sustainable cosmetic products?

From the obtained result, it can be stated that less than half of the respondents (37.7%) have had the experience of buying sustainable cosmetic products, while 32 (46.4%) were uncertain about it, hence it would give reliable results of the research for its objectives.

Influence of Green Marketing on buying of Green Cosmetic Products

To know whether Green Marketing has any significant impact on the buying of green cosmetic products, ANOVA and T-test analysis are used. Green Marketing is a factor and buying green cosmetics is a dependent variable. The different independent variables used for Green Marketing are: -

I get attracted by: -

- V1: Stores with a fresh and vibrant boutique look (sustainable furniture)
- V2: Stores with LED lighting and non-toxic paints
- V3: Recycle packaging
- V4: Environment campaigns
- V5: Community building and giving back to society

The exact significance level (*p* value) is 0.014 as displayed in the sixth Col. (Sig.) of table ANOVA output for Green Marketing. Since *p* value is less than the α value of 0.05, the null hypothesis is rejected. Hence, green marketing impacts the consumer's choice of green cosmetic products. The sixth column of table T-test analysis for Green Marketing shows the

Table 1: ANOVA Output for Green Marketing.

Model	Sum of Squares	df	Mean Square	F	Sig.
Regression	11.383	5	2.277	3.111	.014[b]
Residual	46.096	63	0.732		
Total	57.478	68			

a. Dependent Variable: Have you ever bought any sustainable cosmetic products?
b. Predictors: (Constant), Community building and giving back to society, Stores with a fresh and vibrant boutique look (sustainable furniture), Stores with LED lighting and non-toxic paints, Recycle packaging, Environment campaigns

Table 2: T-test Analysis for Green Marketing.

Coefficients					
Model	Unstandardized Coefficients		Standardized Coefficients	t	Sig.
	B	Std. Error	Beta		
(Constant)	1.821	.310		5.871	.000
Stores with a fresh and vibrant boutique look (sustainable furniture)	.284	.136	.258	2.090	.041
Recycle packaging	−.371	.165	−.297	−2.249	.028
Stores with LED lighting and non-toxic paints	.299	.137	.287	2.178	.033
Environment campaigns	−.130	.186	−.095	−.698	.488
Community building and giving	−.106	.200	−.075	−.528	.600

a. Dependent Variable: Have you ever bought any sustainable cosmetic products?

individual significance values of five independent variables. Out of these five variables, v4 and v5 have p value more than the α value so there is no effect on them the dependent variable, but v1, v2 and v3 significantly impact the dependent variable as its p value is less than the α value. Based on the obtained results, there exists a positive relationship between green consumerism and consumer choice of green cosmetic products. But out of the five factors taken into consideration for determining the significance of social status, only the factors 'Stores with a fresh and vibrant boutique look (sustainable furniture)', 'Recycle packaging 'and 'Stores with LED lighting and non-toxic paints' have a significance towards green consumerism.

Relationship between Green Consumerism and Social Status

To know whether there exists any significant relationship between Green Consumerism of cosmetic products and Social Status, ANOVA and T-test analysis are used. The parameters taken for social status are knowledge, prestige, and wealth. The focus is on the perception of one's social standing relating to others in society. Social Status is a factor and buying green cosmetics is a dependent variable. The different independent variables used for Social Status are: -

- V1: Buying green cosmetic products helps me express my personality.
- V2: I am willing to sacrifice luxury and performance by buying green cosmetic products, if.
- V3: Buying green cosmetic products will make me feel like a responsible citizen.
- V4: Buying green cosmetic products will make me look knowledgeable.
- V5: Buying green cosmetic products will make me look wealthy.
- V6: Buying green cosmetic products will make me look ethical.
- V7: You can tell a lot about a person from whether he/she buys green cosmetic products.

The exact significance level (p value) is .005 as displayed in the sixth Col. (Sig.) of table ANOVA output for Social Status. Since p value is less than α value of 0.05, the null hypothesis is rejected. Hence, green consumerism does lead to a sign of high social status. The sixth column. of table Analysis for Social Status shows the individual significance values of the independent variables. Out of these variables, except v2, all of them have p value more than α value so there is no effect on them the dependent

Table 3: ANOVA Output for Social Status.

Model	Sum of Squares	df	Mean Square	F	Sig.
Regression	14.595	6	2.433	3.517	.005[b]
Residual	42.883	62	.692		
Total	57.478	68			

a. Dependent Variable: Have you ever bought any sustainable cosmetic products?
b. Predictors: (Constant), You can tell a lot about a person from whether he/she buys green cosmetic products., Buying green cosmetic products will make me feel as a responsible citizen., I am willing to sacrifice luxury and performance by buying green cosmetic products, if., Buying green cosmetic products will make me look knowledgeable., Buying green cosmetic products will make me look ethical., Buying green cosmetic products will make me look wealthy.

Table 4: T-test Analysis for Social Status.

Coefficients					
Model	Unstandardized Coefficients		Standardized Coefficients	t	Sig.
	B	Std. Error	Beta		
(Constant)	1.251	.290		4.312	.000
I am willing to sacrifice luxury and performance by buying green cosmetic products, if.	−.077	.072	−.128	−1.070	.289
Buying green cosmetic products will make me feel as a responsible citizen.	.264	.080	.383	3.315	.002
Buying green cosmetic products will make me look knowledgeable.	.139	.095	.201	1.460	.149
Buying green cosmetic products will make me look wealthy.	−.095	.109	−.130	−.873	.386
Buying green cosmetic products will make me look ethical.	−.090	.085	−.153	−1.058	.294
You can tell a lot about a person from whether he/she buys green cosmetic products.	.162	.098	.272	1.656	.103

a. Dependent Variable: Have you ever bought any sustainable cosmetic products?

variable. So, v2 significantly impacts the dependent variable as its p value is less than the α value. Based on the result obtained, there exists a positive relationship between green consumerism and social status. Among the five factors considered for determining the significance of green marketing, only the factor, 'Buying green cosmetic products will make me feel like a responsible citizen, has significance and the rest factors have no significance on the social status.

Influence of Psychographic factors on Buying of Green Cosmetic Products

This section presents the psychographic factors that influence on buying of Green Cosmetic Products. ANOVA test is used to know whether the brand has a significant impact on the purchase behaviour of green cosmetics.

Table 5: Psychographic factors Analysis for Green Cosmetic Products.

Variables	Eta2	F	Sig.
V1: Sustainability	0.331	0.388	0.535
V2: Safety and Quality	0.237	0.277	0.600
V3: Brand connection	0.331	0.338	0.432
V4: Pricing	0.689	0.813	0.370
V5: High product involvement	0.739	1.163	0.101

13.5 Conclusion

The findings of this study provide marketers with some insights into how they should continue to lead and take the necessary steps to remove the present hurdles to green cosmetic product acceptance. People are aware of green cosmetic product acceptance. People are aware of green cosmetics goods, according to the data, but they believe there is less information about them in marketing communications. As a result, effective green cosmetics branding tactics should be established in order to increase customer trust in green goods and advertisements. Furthermore, the correct display of green cosmetic items in a store should be considered, as it is one of the most important aspects in capturing a customer's attention. As a result, the next step should be to develop accurate knowledge about these goods and services, and then ensure that these goods are widely available in the marketplace. Marketers should also visit rural areas and begin raising consciousness as a first move, as rural residents are likely to be less aware of the benefits of using green goods than urban residents. In order to reach buyers of green cosmetic products directly, effective demographic and psychographic segmentation should be implemented, bearing in mind the factors that have the biggest effect on consumer purchasing of green cosmetic products.

References

Amberg, N., and C. Fogarassy. 2019. Green consumer behavior in the cosmetics market. *Resources* 8(3): 137.

Askadilla, W. L., and M. N. Krisjanti. 2017. Understanding indonesian green consumer behavior on cosmetic products: Theory of planned behavior model. *Polish Journal of Management Studies* 15.

Born, C. T., S. E. Ross, W. M. Iannacone, C. W. Schwab, and W. G. DeLong. 1989. Delayed identification of skeletal injury in multisystem trauma: the'missed'fracture. *The Journal of Trauma* 29(12): 1643–1646.

Fonseca-Santos, B., M. A. Corrêa, and M. Chorilli. 2015. Sustainability, natural and organic cosmetics: consumer, products, efficacy, toxicological and regulatory considerations. *Brazilian Journal of Pharmaceutical Sciences* 51: 17–26.

Ghazali, E., P. C. Soon, D. S. Mutum, and B. Nguyen. 2017. Health and cosmetics: Investigating consumers' values for buying organic personal care products. *Journal of Retailing and Consumer Services* 39: 154–163.

Rawat, S. R., and P. Garga. 2012. Understanding consumer behaviour towards green cosmetics. *Available at SSRN 2111545*.

Sadiq, Mohd, M. Adil, and J. Paul. 2021. An innovation resistance theory perspective on purchase of eco-friendly cosmetics. *Journal of Retailing and Consumer Services* 59: 102369.

Sharma, M., P. Trivedi, and J. Deka. 2021. A paradigm shift in consumer behaviour towards green cosmetics: An empirical study. *International Journal of Green Economics* 15(1): 1–19.

Sreen, N., A. Dhir, S. Talwar, T. M. Tan, and F. Alharbi. 2021. Behavioral reasoning perspectives to brand love toward natural products: Moderating role of environmental concern and household size. *Journal of Retailing and Consumer Services* 61: 102549.

Sousa, Bruno, Veloso, Claúdia and Magueta, Daniel. 2020. Green marketing as an instrument for business competitiveness: A theoretical contribution. 35th IBIMA Conference: 1–2 April 2020, (pp. 9384–9393) Seville, Spain.

Testa, F., N. Gusmerotti, F. Corsini, and E. Bartoletti. 2022. The role of consumer trade-offs in limiting the transition towards circular economy: The case of brand and plastic concern. *Resources, Conservation and Recycling* 181: 106262.

Tewari, A., S. Mathur, S. Srivastava, and D. Gangwar. 2022. Examining the role of receptivity to green communication, altruism and openness to change on young consumers' intention to purchase green apparel: A multi-analytical approach. *Journal of Retailing and Consumer Services* 66: 102938.

CHAPTER 14

Analytical Study of Factors Affecting the Adoption of Blockchain by Fintech Companies

Pooja Mathur* and Sony Thakural

Contents

14.1 Introduction
14.2 Background of Blockchain Technology
14.3 Literature Review
14.4 Research Methodology
14.5 Data Analysis
14.6 Conclusion

14.1 Introduction

"Change is the only constant", the rate of change is ever increasing. Undoubtedly everything in nature is bound to modify itself. Geographically, climatically, and technologically we have all witnessed the ameliorations that have taken place in the history of mankind. Sometimes 'change' brings with it, extreme uncertainty. All the changes that the business world witnesses have their connotations when viewed from the angles of marketing, social welfare and consumer behavior. Blockchain has emerged as a digital technology that seeks to ensure that financial transactions are carried out in the arena of security and transparency.

Amity School of Business, Amity University, Uttar Pradesh, Noida.
* Corresponding author: poojamathur6111986@gmail.com

The impressive amelioration in the FinTech industry has simulated the adoption of blockchain technology and associated cryptocurrencies.

Whenever a business adopts a new kind of technology it comes with its set of implications that affect not only the Information Technology department of the organization but have some trickle-down effect on all business processes and functions. Nonetheless, the influence of technologies is different and so is the adoption process that fintech companies tend to follow. Quite recently, the amelioration in technology has brought the revolution of Blockchain into the ecosystem of the global village. Firms must understand the dimensions of the adoption process throughout the organization and also acquaint themselves with the decision-making factors about these dimensions. It is important at this stage to understand the basics of Blockchain technology and how it impacts the future of the financial world. In connection with the topic, an effort to link Blockchain technology with the organization's approach to its adoption shall be made.

Blockchain has established a stance for itself by way of managing transactions in real-time across the boundaries of enterprises and stimulating collaboration in business endeavors. The pivotal advantage of blockchain depends on its ability to reduce transaction management costs to a great extent. Blockchain has been able to develop a very holistic approach by catering to a variety of sectors including healthcare, education, social media, and the entertainment industry. Stejskal et al. (2019) mentioned in their report the pervasiveness of blockchain and how it amplifies knowledge creation and an innovative approach through technological deployment.

14.2 Background of Blockchain Technology

Blockchain technology for the first time was elucidated by Nakamoto in the year 2008 to be related to a computer architecture mechanism that enables sending digital instructions through a well-regulated mechanism over the internet. There are built-in nodes that execute digital transactions and maintain a record of the same. All these transactions are further stacked together into blocks, which rightfully gives the technology its name. These blocks efficaciously form a series of digital instructions that are woven together in a chain which in turn ensures that the content and chronology of the transactions remain authenticated. Ozilli (2018) has described blockchain as a digital ledger where a record of transactions is kept in an open source and chronological format. No single entity has control or autonomy over the data and information comprising the database. The financial services industry has benefited from the use of Blockchain which is often called "Distributed Ledger Technology". According to a report by

Infosys, it is through Blockchain that financial institutions have witnessed savings in their infrastructure, transactions, and various administrative costs. The very fact that the mechanism of blockchain eliminates the involvement of any centralized entity, the digital transfer of financial assets has been disintermediated to a great extent. The financial ecosystem has become more trustworthy and sturdier with the deployment of blockchain. Santander in one of its reports rolled out in 2015 mentioned that with proper implementation of blockchain by the year 2022 banks could reduce their infrastructure costs regarding securities trading, regulatory compliance, and cross-border payment-related aspects.

Blockchain is most used for trading purposes related to cryptocurrency in the form of bitcoins. As far as crypto blockchains are concerned the buying and selling of bitcoins or any other type of cryptocurrency is recorded in a delineated and chronological format. Several users can see all the transactions in a blockchain pattern and have the right to add new sets of information to the chain of existing transactions in the blockchain network. The protocol of joining a blockchain varies from network to network, hence some make it compulsory for the user to seek permission to join a specific network.

14.3 Literature Review

Every researcher must develop a thorough understanding of all the previous research about a topic or a body of knowledge that came before. Literature Review helps a researcher in giving a 360° approach to his work by elucidating the studies that other researchers have conducted in the past, on either the same topic or some other closely associated one.

Numerous individuals have worked on research papers that deal with the adoption of Fintech and Blockchain in the financial sector. The efforts in Chapter two are channelized toward documenting some of these so that the reader of this work can perceive the different facets of the theoretical arena concerned with our topic of discussion.

World Economic Forum in its 2016 report mentioned that more than 24 countries have planted investments in distributed ledger technology which is also known as blockchain technology. Along with that more than 90 federal banks along with major corporations have become a part of the blockchain consortia.

Yurcan (2016) elaborated on how Deloitte, one of the big 4 firms realized the potential that blockchain holds. It got into a partnership with almost five firms that have a specialization in technology to leverage the benefits of blockchain in the consulting business and subsequently have come up with blockchain-based applications like digital identities and cross-border payment mechanisms.

Scott et al. (2016) enumerated how Capital markets constitute an important part of the larger financial system and house numerous financial instruments like securities, bond instruments and other long-term horizon investments to appreciate the capital of companies. These stages cover pre-trade, trade, post-trade and finally custody and securities servicing. The following image summarizes the applications of blockchain during various stages of trading in the financial markets.

Roman Beck et al. (2016) mentioned that blockchain technology is secure to a great extent but has certain scalability issues, cost-related factors and a degree of volatility in transaction currency are some of the issues that come up. Matilla (2016) elucidated on how Distributed Ledger Technology became fuel for a wide variety of applications.

Walsh et al. (2016) mention in their report how researchers have always considered the technical aspect of blockchain technology and as a whole the business and the user-oriented perspectives are absent. This is a very important aspect and has served as an indicator of the gap in research.

Ebers and Maurer (2016) believe that employees need a sense of validation from the top management before they adopt any new technology, hence companies should try to carry out proper adoption decision processes for a seamless rollout of innovations especially those related to blockchain.

Glaser (2017) writes about how the present blockchain-related phenomena are innovative but are still on the lookout for potential uses and deployment possibilities. He also describes how the research arena as well as practical implementation have a paucity of effective approaches through which better applications can be developed.

Risius and Spohrer (2017) have also highlighted the aspect that blockchain technology holds a lot of potentials but there is a severe lack of a body of knowledge that can discreetly enlist all possible uses.

Morabito (2017) has described how numerous organizations realized the benefits of the timely adoption of blockchain technology. The author has given an example of IBM and how it joined the Chamber of Digital Commerce that came into being with the coming together of some blockchain startups, software companies, investors and financial institutions to work in a close-knit network with the American Government and also chalk out standards for developing blockchain technology and its usage. It also considered the legal aspects related to the adoption of blockchain.

Khrais (2017) described the business infrastructure to have become impressively digitized which is good news for blockchain technology to become a success and dominate the business world.

Roman Beck et al. (2017) investigated the affirmative implications of blockchain technology for present-day organizations typically those that are in the financial services industry or tangible asset ownership. The chapter seeks to highlight the broader perspective on using blockchain technology. From handling refined financial transactions to dealing with money transfers that transcend geographical boundaries, blockchain has made a very impressive mark in the business world. The DLT keeping a tab on the ownership of a variety of assets, according to authors is a very valuable application of the technology. They also mention how various intermediaries like traders, financers, insurers, and other regulators can have their needs and goals met only through the deployment of blockchain technology.

Rohmat et al. (2018) sought to chalk out the determinants of blockchain technology adoption of defined payment systems in the Indonesian banking industry. The researchers developed a novel model of the payment system by making intelligent use of blockchain technology. For quite some time the research on blockchain technology had been going on in developed countries like the UK, America, Germany, Japan, Scotland, France, and China but not much had been done in the context of developing countries. The researchers identified 9 factors that are directly linked to the adoption of blockchain technology in the context of payment systems. Secondly, a payment system mechanism model by making use of blockchain was also introduced that was made easy to use, ensured privacy, a very meagre number of transactions and also speedy and efficacious remittances. These advantages will also contribute to the competitive stance of the Indonesian Banking industry.

Ku-Mahmud et al. (2018) carried out a study to assess the growth in the arena of blockchain, cryptocurrency and the overall Fintech market dynamics. A survey was done on 304 blockchain users and other industry-related entities in Malaysia. As per the results, privacy was seen as the most significant obstacle to the seamless adoption of blockchain technology and its associated cryptocurrency aspect. Legal perspective, the hefty cost of deployment, security, and difficulties in the context of switching technology were seen as prime barriers. The researchers have also stated that the banking sector is the most influential in the Fintech industry. Hence, business decision-making entities need to timely evaluate the probable application of blockchain in their line of business and accordingly take strategic decisions.

Gimpel et al. (2018) mention how upcoming digital technology systems act as disruptors and make the market strive for the better and change with time. This has brought about a dynamic reengineering in business practices and the status quo.

Panarello et al. (2018) state that a rise in the number of users has caused blockchain to go through an issue of scalability. The response time and cost are two factors that change tremendously because of issues of scalability. The increased traffic that comes up because of a larger user influx can only be dealt with by adding more nodes to the network to increase the processing speed. This, however, will add to the cost. A comprehensive solution is needed to overcome the issue of scalability.

Tatjana Boshkov (2019) has chalked out an analysis of the applications of cryptocurrency and especially bitcoin. Financial and data privacy have been addressed in this chapter too as they are extremely sensitive issues and building a safe payment system using standardization, compatibility and a tremendous amount of innovation.

Omar et al. (2019) researched to investigate the awareness of blockchain in Malaysia. Other factors considered were trust and the overall adoption perspective of blockchain. Awareness, trust, and adoption were first checked with the help of a questionnaire. The authors sought to use both descriptive as well as inferential statistics to analyze the results. The authors have mentioned that the future of blockchain in Malaysia could be overshadowed due to the government's excessive regulations. The lack of guidelines about blockchain deployment can pose serious challenges in the adoption of technology, especially in Malaysian financial institutions.

Holotuik and Moormann (2019) have taken the humanistic side of blockchain adoption in companies by stating that the successful implementation and adoption of blockchain technology depends on the human capital of the organization and its caliber concerning its ability to use blockchain in various business functions and IT sub-units. Thus, before the adoption is done, the capable individuals should be identified as they can serve as enablers and facilitators of blockchain adoption in any company.

As per an article published online by Economic Times in February 2020, "Trust in tech firms, Demand for banking, new market players in the fintech domain, increased focus on security and privacy, chatbots that are backed by NLP, cloud banking, insurance technology along with peer-to-peer lending mechanisms" are the eight factors that strongly drive blockchain adoption.

Khurshid (2020) has taken a very novel perspective in analyzing the implications of the COVID-19 pandemic on Blockchain technology and its applications. The significantly high number of casualties during the COVID-19 pandemic has highlighted the absence of sturdy medical infrastructure and why human health and welfare should be prioritized. The objective of the research chapter is mainly to elucidate the concept of blockchain and cryptography-related security and how these two can provide helpful solutions to trust problems about data. The chapter describes how blockchain technology can help in tracking the medical supplies of infected individuals. The privacy-ensuring features of blockchain technology can be leveraged to create systems that can solve the contention between data privacy and addressing the needs related to public health, especially in the context of the pandemic. The author has mentioned how a coordinated endeavor on a national level could address the present lacuna in the healthcare sector and fill it up with the potential application of blockchain technology and the dedicated contribution of academia, business and commerce and research domain could speed up adoption blockchain, especially in the healthcare sector.

14.4 Research Methodology

The research work consists of a theoretical elucidation of major concepts about blockchain technology and its applications. All efforts to develop a basic understanding and build up a model based on the user's perspective of concerns related to the privacy of data and their preference for Robo advisors over human advisors and how these two aspects are related to the impact of blockchain in three different sectors have been made. The main objective of the study is to critically understand whether a statistically significant relationship exists between Data Privacy and User Preference of a Robo vs. Human Advisor with the impact of blockchain technology on Commercial Banking, Real Estate and Capital Markets.

Primary research in this proposed work will be carried out with the help of a structured questionnaire which had been circulated online to several individuals who possess a basic understanding of blockchain technology. The sampling technique followed is snowball sampling. Carlos (2021) has also adopted a similar sampling technique in his study. The number of respondents is 102. The Total number of questions in the questionnaire is 10, out of which 5 questions seek to obtain basic demographic information. The 5 questions sought to enquire the perspective of the respondents concerning the usage of e-wallets, their understanding of the impact of blockchain, the regulatory compliance aspect and concerns related to cyber security with the help of Likert scale statements.

The Analysis of the responses is carried out by applying Descriptive Statistics as well as Inferential Statistics. Hypothesis testing has been carried out in Inferential Statistics.

14.5 Data Analysis

In this study, the ordinal logistic regression model has been utilized for the analysis of the collected data. The logistical model of regression came because of the incessant hard work Mr. Joseph Berkson carried out for over a decade. It is viewed as an alternative to the probit model wherein the dependent variable is limited to only two values. The logit model developed by Berkson was only considered inferior to the probit model but gradually became a very considerable model of analysis also transcending the probit model. This model is regarded as being simplistic in nature as far as computation is concerned. Its mathematical properties pave way for its usage in numerous arenas and fields. David Cox in his research work published in the year 1958 recommended certain changes to the model to make it more refined. Furthermore, in the years 1966 and 1969, Cox and Thiel respectively introduced the multinomial model, which ameliorated the scope of application of the logistical model of regression and made it more renowned amongst statisticians. McFadden in the year 1973 gave us the logistic model of regression, a theoretical foundation.

HYPOTHESIS TESTING-I

Null Hypothesis I(a): The risk of an absence of data privacy has an insignificant statistical relationship with the impact of blockchain technology on Commercial Banking.

Null Hypothesis I(b): User Preference for a Robo advisor over a human advisor has an insignificant statistical relationship with the impact of blockchain technology on Commercial Banking.

Ordinal Logistic Regression Results—Commercial Banking

Table 1: Case Processing Summary.

		N	Marginal Percentage
Impact Commercial Banking	Disagree	1	1.0%
	Neutral	6	5.9%
	Agree	8	7.8%
	Strongly Agree	87	85.3%
Valid		102	100.0%
Missing		0	
Total		102	

Table 2: Model Fitting Information.

Model	-2 Log Likelihood	Chi-Square	df	Sig.
Intercept Only	28.586	2.621	2	.270
Final	25.966			

Link function: Logit.

The model fitting information table gives us an account of the p value which is 0.270 and must be compared with the significance value of 0.05. Since the p value is > 0.05 which means [0.270 > 0.05] that means that the full model does not statistically significantly predict the dependent variable better than the intercept-only model which is the regression model without predictors.

Table 3: Goodness-of-Fit.

	Chi-Square	Df	Sig.
Pearson	6.448	13	.928
Deviance	6.954	13	.904

Link function: Logit.

To check the overall measure of the model and if our model fits the data well, we will have to compare the p value with the significance level of 0.05. The p value corresponding to the Pearson's coefficient with a degree of freedom 13, is 0.928 which is greater than 0.05, hence a non-significant result in Goodness of fit means that the model fits the data well.

Table 4: Pseudo R-Square.

Cox and Snell	.025
Nagelkerke	.038
McFadden	.023

Link function: Logit

In the Pseudo R-Square table given above, there is no fixed guideline as to which reading, we shall prefer from the three given above but as a common practice amongst researchers, the Nagelkerke pseudo-r-square value is chosen which in this case is 0.038 which basically means that 3.8% variability brought by independent variables on the dependent variable.

The risk of absence of data privacy and User Preference of a Robo advisor over a human advisor introduces a variability of 3.8% on the impact of blockchain technology on Commercial Banking.

Table 5: Test of Parallel Linesa.

Model	-2 Log Likelihood	Chi-Square	df	Sig.
Null Hypothesis	25.966			
General	25.152	.814	4	.937

The null hypothesis states that the location parameters (slope coefficients) are the same across response categories.

a. Link function: Logit.

The test of parallel lines tests the assumption of proportional odds. To be able to apply the Ordinal Logistic regression we need to have a p value > 0.05. If the result of the test of parallel lines indicates non-significance, then we interpret it to mean that the assumption stands satisfied and if the opposite is true as in the case of a significant result, then the assumption is violated.

In this case, we have fulfilled the assumption and can easily interpret the result from the parameter estimates. As p value of 0.937 is greater than the significance value of 0.05.

Table 6: Parameter Estimates.

		Estimate	Std. Error	Wald	df	Sig.	95% Confidence Interval	
							Lower Bound	Upper Bound
Threshold	[Impact CommercialBanking = 2.00]	−64.826	1.418	2089.914	1	.000	−67.605	−62.047
	[Impact CommercialBanking = 3.00]	−62.804	1.085	3353.114	1	.000	−64.930	−60.679
	[Impact CommercialBanking = 4.00]	−61.940	1.061	3410.410	1	.000	−64.019	−59.861
Location	DataPrivacy	−12.325	.000	-	1	-	−12.325	−12.325
	AutomatedAdvisory	.341	.256	1.772	1	.183	−.161	.844

Link function: Logit.

· The parameter estimates table gives the regression coefficients and significance tests for the independent variable in the model. The regression coefficients will be interpreted as the predicted change in the log odds of being in a higher or lower category on the dependent variable per unit increase or decrease in the independent variables.

· The risk of absence of Data Privacy is not a significant predictor in the model. The coefficient can be interpreted as: For every one unit increase in the risk of absence of data privacy, there is a predicted decrease of 12.325 in the log odds of being in a higher level of impact of blockchain on commercial banking.

· In the estimates column, we see that Automated Advisory which reflects the User Preference of a Robo over Human Advisor shows a positive estimate of 0.341 which makes it a significant positive predictor of the impact of blockchain technology on Commercial Banking. For every one unit increase in user preference for automated advisory variable, there is a predicted increase of 0.341 in the log odds of the impact of blockchain on commercial banking being in a higher category.

Results and Conclusion for Hypothesis Testing-I

· Based on the results obtained, we see that risk of the absence of data privacy is not a significant predictor of the impact of blockchain on commercial banking which the Null Hypothesis I(a) stated. Hence in this case Null Hypothesis I(a) cannot be rejected.

· Based on the results obtained, we see that the User Preference for a Robo over a Human Advisor is a significant positive predictor of the impact of blockchain technology on Commercial Banking which Alternative Hypothesis I(b) stated. Hence in this case Alternative Hypothesis I(b) can be supported.

HYPOTHESIS TESTING-II

Null Hypothesis II(c): The risk of absence of data privacy has an insignificant statistical relationship with the impact of blockchain technology on Real Estate.

Null Hypothesis II(d): User Preference for a Robo advisor over a human advisor has an insignificant statistical relationship with the impact of blockchain.

Ordinal Logistic Regression Results—Real Estate

Table 7: Case Processing Summary.

		N	Marginal Percentage
Impact Real Estate	Strongly Disagree	7	6.9%
	Disagree	17	16.7%
	Neutral	14	13.7%
	Agree	25	24.5%
	Strongly Agree	39	38.2%
Valid		102	100.0%
Missing		0	
Total		102	

Table 8: Model Fitting Information.

Model	-2 Log Likelihood	Chi-Square	df	Sig.
Intercept Only	71.259	1.890	2	.389
Final	69.368			

Link function: Logit.

The model fitting information table gives us an account of the p value which is 0.389 and must be compared with the significance value of 0.05. Since p value > 0.05 which means [0.389 > 0.05] and means that the full model does not statistically significantly predict the dependent variable better than the intercept-only model which is the regression model without predictors.

Table 9: Goodness-of-Fit.

	Chi-Square	df	Sig.
Pearson	29.021	18	.048
Deviance	29.166	18	.046

Link function: Logit.

To check the overall measure of the model and if our model fits the data well, we will have to compare the p value with the significance level of 0.05. The p value corresponding to Pearson's coefficient with a degree of freedom of 18, is 0.048 which is lesser than 0.05, hence a statistically significant result in Goodness of fit means that the model does not fit the data well.

Table 10: Pseudo R-Square.

Cox and Snell	.018
Nagelkerke	.019
McFadden	.006

Link function: Logit

In the Pseudo R-Square table given above, there is no fixed guideline as to which reading, we shall prefer from the three given above but as a common practice amongst researchers, the Nagelkerke pseudo-r-square value is chosen which in this case is 0.019 which basically means that 1.9% variability brought by independent variables on the dependent variable.

The risk of an absence of data privacy and User Preference of a Robo advisor over a human advisor introduce variability of 1.9% on the impact of blockchain technology on Real Estate.

Table 11: Test of Parallel Linesa.

Model	-2 Log Likelihood	Chi-Square	df	Sig.
Null Hypothesis	69.368	8.133	6	.229
General	61.235			

The null hypothesis states that the location parameters (slope coefficients) are the same across response categories.

a. Link function: Logit.

The test of parallel lines tests the assumption of proportional odds. To be able to apply the Ordinal Logistic regression we need to have a p value > 0.05. If the result of the test of parallel lines indicates non-significance, then we interpret it to mean that the assumption stands satisfied and if the opposite is true as in the case of a significant result, then the assumption is violated.

In this case, we have fulfilled the assumption and can easily interpret the result from the parameter estimates. As p value of 0.229 is greater than the significance value of 0.05.

Table 12: Parameter Estimates.

		Estimate	Std. Error	Wald	df	Sig.	95% Confidence Interval	
							Lower Bound	Upper Bound
Threshold	[ImpactRealEstate = 1.00]	2.607	5.209	.251	1	.617	−7.601	12.816
	[ImpactRealEstate = 2.00]	4.039	5.213	.600	1	.438	−6.178	14.255
	[ImpactRealEstate = 3.00]	4.707	5.219	.813	1	.367	−5.522	14.935
	[ImpactRealEstate = 4.00]	5.731	5.228	1.202	1	.273	−4.516	15.978
Location	DataPrivacy	.911	1.043	.762	1	.383	−1.135	2.956
	AutomatedAdvisory	.171	.182	.880	1	.348	−.186	.528

Link function: Logit.

· The parameter estimates table gives the regression coefficients and significance tests for the independent variable in the model. The regression coefficients will be interpreted as the predicted change in the log odds of being in a higher or lower category on the dependent variable per unit increase or decrease in the independent variables.

· The risk of absence of Data Privacy is a significant predictor in the model. The coefficient can be interpreted as: For every one unit increase in the risk of absence of data privacy, there is a predicted increase of 0.911 in the log odds of being in a higher level of impact of blockchain on real estate.

· In the estimates column, we see that Automated Advisory which reflects the User Preference of a Robo over Human Advisor shows a positive estimate of 0.171 which makes it a significant positive predictor of the impact of blockchain technology on Real Estate. For every one unit increase in user preference for automated advisory variable, there is a predicted increase of 0.171 in the log odds of the impact of blockchain on real estate being in a higher category.

Results and Conclusion for Hypothesis Testing-II

· Based on the results obtained, we see that risk of absence of data privacy is a significant predictor of the impact of blockchain on real estate which Alternative Hypothesis II(c) stated. Hence in this case Alternative Hypothesis II(c) can be supported.

· Based on the results obtained, we see that the User Preference of a Robo over a Human Advisor is a significant positive predictor of the impact of blockchain technology on Real Estate which the Alternative Hypothesis II(d) stated. Hence in this case Alternative Hypothesis II(d) can be supported.

HYPOTHESIS TESTING-III

Null Hypothesis III(e): The risk of absence of data privacy has an insignificant statistical relationship with the impact of blockchain technology on Capital Markets Infrastructure.

Null Hypothesis III(f): User Preference for a Robo advisor over a human advisor has an insignificant statistical relationship with the impact of blockchain technology on Capital Markets Infrastructure.

Ordinal Logistic Regression Results—Capital Market Infrastructure

Table 13: Case Processing Summary

		N	Marginal Percentage
Impact Capital Markets	Strongly Disagree	3	2.9%
	Disagree	5	4.9%
	Neutral	12	11.8%
	Agree	19	18.6%
	Strongly Agree	63	61.8%
Valid		102	100.0%
Missing		0	
Total		102	

Table 14: Model Fitting Information.

Model	-2 Log Likelihood	Chi-Square	df	Sig.
Intercept Only	52.547			
Final	50.486	2.061	2	.357

Link function: Logit.

The model fitting information table gives us an account of the p value which is 0.357 must be compared with the significance value of 0.05. Since p value > 0.05 which means [0.357 > 0.05] that means that the full model does not statistically significantly predict the dependent variable better than the intercept-only model which is the regression model without predictors.

Table 15: Goodness-of-Fit.

	Chi-Square	df	Sig.
Pearson	15.384	18	.635
Deviance	17.377	18	.497

Link function: Logit.

To check the overall measure of the model and if our model fits the data well, we will have to compare the p value with the significance level of 0.05. The p value corresponding to Pearson's coefficient with a degree of freedom of 18, is 0.635 which is greater than 0.05, hence a non-significant result in Goodness of fit means that the model fits the data well.

Table 16: Pseudo R-Square.

Cox and Snell	.020
Nagelkerke	.022
McFadden	.009

Link function: Logit

In the Pseudo R-Square table given above, there is no fixed guideline as to which reading, we shall prefer from the three given above but as a common practice amongst researchers, the Nagelkerke pseudo-r-square value is chosen which in this case is 0.022 which means that 2.2% variability brought by independent variables on the dependent variable.

Risk of the absence of data privacy and User Preference of a Robo advisor over a human advisor introduce variability of 2.2% on the impact of blockchain technology on Capital Markets Infrastructure.

Table 17: Test of Parallel Linesa.

Model	-2 Log Likelihood	Chi-Square	df	Sig.
Null Hypothesis	50.486	16.944c	6	.149
General	33.542b			

The null hypothesis states that the location parameters (slope coefficients) are the same across response categories.

a. Link function: Logit.

The test of parallel lines tests the assumption of proportional odds. To be able to apply the Ordinal Logistic regression we need to have a p value > 0.05. If the result of the test of parallel lines indicates non-significance, then we interpret it to mean that the assumption stands satisfied and if the opposite is true as in the case of a significant result, then the assumption is violated.

In this case, we have fulfilled the assumption and can easily interpret the result from the parameter estimates. As p value of 0.149 is greater than the significance value of 0.05.

Table 18: Parameter Estimates.

		Estimate	Std. Error	Wald	df	Sig.	95% Confidence Interval	
							Lower Bound	Upper Bound
Threshold	[Impact CapitalMarkets = 1.00]	2.748	5.300	.269	1	.604	−7.641	13.137
	[Impact CapitalMarkets = 2.00]	3.785	5.288	.512	1	.474	−6.579	14.150
	[Impact CapitalMarkets = 3.00]	4.851	5.295	.839	1	.360	−5.526	15.228
	[Impact CapitalMarkets = 4.00]	5.802	5.306	1.195	1	.274	−4.598	16.202
Location	DataPrivacy	1.102	1.059	1.083	1	.298	−.974	3.178
	AutomatedAdvisory	.193	.198	.947	1	.330	−.195	.580

Link function: Logit.

• The parameter estimates table gives us the regression coefficients and significance tests for the independent variable in the model. The regression coefficients will be interpreted as the predicted change in the log odds of being in a higher or lower category on the dependent variable per unit increase or decrease in the independent variables.

· The risk of the absence of Data Privacy is a significant positive predictor in the model. The coefficient can be interpreted as: For every one unit increase in the risk of absence of data privacy, there is a predicted increase of 1.102 in the log odds of being in a higher level of impact of blockchain on capital market infrastructure.

· In the estimates column, we see that Automated Advisory which reflects the User Preference of a Robo over Human Advisor shows a positive estimate of 0.193 which makes it a significant positive predictor of the impact of blockchain technology on Capital Market Infrastructure. For every one unit increase in user preference for automated advisory variable, there is a predicted increase of 0.193 in the log odds of the impact of blockchain on capital market infrastructure being in a higher category.

Results and Conclusion for Hypothesis Testing-III

· Based on the results obtained, we see that risk of absence of data privacy is a significant positive predictor of the impact of blockchain technology on capital market infrastructure which Alternative Hypothesis III(e) stated. Hence in this case Alternative Hypothesis III(e) can be supported.

· Based on the results obtained, we see that the User Preference for a Robo over a Human Advisor is a significant positive predictor of the impact of blockchain technology on Capital market infrastructure which Alternative Hypothesis III(f) states. Hence in this case Alternative Hypothesis III(f) can be supported.

14.6 Conclusion

The adoption of Blockchain by Fintech companies has redefined itself in many ways over the years. The investment in blockchain technology is directly linked to the various sectors that see lucrative business opportunities because of deploying distributed ledger technology. Fintech Companies have emerged as the major service providers who refine the experience of their clients by leveraging upon the knowledge pool and resources available to them. This Research work has helped one establish several conclusions. The point-wise account of conclusions is as follows: -

1. Blockchain Technology has a long way to go in the financial domain as dedicated research is on to discover the various possibilities of its application.

2. Based on the results obtained, User Preference for a Robo over a Human Advisor is a significant positive predictor of the impact of blockchain technology on Commercial Banking, Real Estate and Capital Markets Infrastructure.

3. Based on the results obtained, we see that risk of absence of data privacy is a significant predictor of the impact of blockchain on real estate and capital markets.

4. The holistic exploratory research summarizes the various factors that determine the adoption of blockchain technology in Fintech companies. The various considerations before the adoption decision are made.

5. Secondary data from open sources is slightly difficult to verify, which is why the authenticity cannot be tested with certainty even for published work.

6. In Primary Research, there is a possibility that the respondents might have answered a question with a biased perspective or without being aware of its actual connotation; such errors are beyond the researcher's control.

References

Beck, R., J. Stenum Czepluch, N. Lollike, and S. Malone. 2016. Blockchain–the gateway to trust-free cryptographic transactions.

Beck, R., M. Avital, M. Rossi, and J. B. Thatcher. 2017. Blockchain Technology in Business and Information Systems Research.

Boshkov, T. 2019. Blockchain and Digital Currency in the World of Finance. Blockchain and Cryptocurrencies, 23.

Deloitte Insights. 2021. Deloitte's 2021 Global Blockchain Survey. [online] Available at: https://www2.deloitte.com/us/en/insights/topics/understanding-blockchain-potential/global-blockchain-survey.html [Accessed 28 October 2021].

Ebers, M., and I. Maurer. 2016. To continue or not to continue? Drivers of recurrent partnering in temporary organizations. *Organization Studies* 37(12): 1861–1895.

ETBFSI.com. 2021. Eight factors that will drive fintech in India this year - ET BFSI. [online] Available at: <https://bfsi.economictimes.indiatimes.com/news/fintech/eight-factors-that-will-drive-fintech-in-india-this- year/74167318> [Accessed 28 October 2021].

Gimpel, H., D. Rau, and M. Röglinger. 2018. Understanding FinTech start-ups–a taxonomy of consumer- oriented service offerings. *Electronic Markets* 28(3): 245–264.

Glaser, F. 2017. Pervasive decentralisation of digital infrastructures: a framework for blockchain enabled system and use case analysis.

Holotiuk, F., and J. Moormann. 2019. Dimensions, Success Factors and Obstacles of the Adoption of Blockchain Technology.

Khrais, L. T. 2017. Framework for measuring the convenience of advanced technology on user perceptions of Internet banking systems. *Journal of Internet Banking and Commerce* 22(3): 1–18.

Ku-Mahamud, K. R., N. A. Abu Bakar, and M. Omar. 2018. Blockchain, cryptocurrency and FinTech market growth in Malaysia. *Journal of Advance Research in Dynamical & Control Systems*, 10(14 SI): 2074–2082.

Ku-Mahamud, K. R., M. Omar, N. A. A. Bakar, and I. D. Muraina. 2019. Awareness, trust, and adoption of blockchain technology and cryptocurrency among blockchain communities in Malaysia. *International Journal on Advanced Science, Engineering and Information Technology* 9(4): 1217–1222.

Khurshid, A. 2020. Applying blockchain technology to address the crisis of trust during the COVID-19 pandemic. *JMIR medical informatics* 8(9): e20477.

Loonam, J., B. Scott, and V. Kumar. 2017. Exploring the rise of Blockchain Technology: Towards Distributed Collaborative Organisations.

Macheel, T. 2016, (June 30). IBM Joins Washington Blockchain Trade Group. Americanbanker. Com. Retrieved October 24, 2021, from https://www.americanbanker.com/bank-technology/ibm-joins-washington-blockchain- trade-group-1081778-1.html

Mattila, J. 2016. The blockchain phenomenon–the disruptive potential of distributed consensus architectures (No. 38). ETLA working papers.

Morabito, V. 2017. Business innovation through blockchain. Cham: Springer International Publishing.

Panarello, A., N. Tapas, G. Merlino, F. Longo, and A. Puliafito. 2018. Blockchain and iot integration: A systematic survey. *Sensors*, 18(8): 2575.

Prokop, V., J. Stejskal, and O. Hudec. 2019. Collaboration for innovation in small CEE countries.

Ramón-Rodríguez, C. L. 2021. Factors Affecting Blockchain Technology Acceptance in Mobile Financial Transactions and Services (Doctoral dissertation, Universidad Ana G Méndez-Gurabo).

Risius, M., and K. Spohrer. 2017. A blockchain research framework. *Business & Information Systems Engineering* 59(6): 385–409.

Scott, A., J. Van De Velde, and I. Dalton. 2016. Blockchain in Capital Markets: the Prize and the Journey.

Taufiq, R., A. N. Hidayanto, and H. Prabowo. 2018, (September). The affecting factors of blockchain technology adoption of payments systems in Indonesia banking industry. In 2018 International Conference on Information Management and Technology (ICIMTech) (pp. 506–510). IEEE.

Vyasa, V. and A. Kumar. 2019. Blockchain Adoption in Financial Services. [online] Infosys. com. Available at: https://www.infosys.com/industries/financial-services/white-papers/documents/blockchain-adoption-financial- services.pdf.

Walsh, C., P. OReilly, R. Gleasure, J. Feller, S. Li, and J. Cristoforo. 2016. New kid on the block: a strategic archetypes approach to understanding the Blockchain.

Yurcan, B. 2016. Blockchain Firms Team Up with Deloitte.

CHAPTER 15

A Study of the Performances of Small Cap, Large Cap and Banking & Financial Sector Funds of Nippon India AMC, ICICI Prudential AMC and Tata AMC

Khushboo Bhasin and *Saloni Pahuja**

Contents

Amity School of Business, Amity University, Uttar Pradesh, Noida.
* Corresponding author: spahuja1@amity.edu

15.1 Introduction

With the constantly changing scenario of the Indian capital market, avenues for investments in financial assets have changed drastically. In past, investors (particularly small investors) had limited investment options, e.g., shares, bonds and debentures, post office deposits, bank FDs, etc. But in recent times investors have got lot more avenues for this purpose and mutual funds are one of them. SEBI (mutual fund) regulations, 1993, defined a mutual fund as, A fund established in the form of a trust by a sponsor to raise money by the trustees through the sale of units to the public under one or more schemes for investing in securities by these regulations. In the current economic era. The Indian mutual fund Industry has emerged as one of the most promising investment opportunities. Investment in financial assets has always been a matter of great importance in an investor's life.

Every investor, no matter how small the savings are, wants to earn a good number of returns at a sustainable rate of risk. To substantiate the diversified financial goals of investors, a variety of mutual fund schemes have surfaced. It is important for both the investors and the fund managers to undergo rigorous and constant evaluation, regarding the risk and return, of various schemes under purview. It enables the fund managers to identify the strengths and weaknesses of these schemes, which helps them to take improved decisions in future.

There are 44 Asset Management Companies (AMCs) or Mutual Fund Houses in India. Every mutual fund institution has funds in the equity, debt, hybrid, and ELSS categories, as well as some in the solution-oriented category. When investors start comparing historical returns and investing their money into them, the problem develops. When the fund does not perform well owing to market conditions, it is unable to provide expected yields to investors, resulting in the problem statement.

About this and to keep on promoting investors' interest in mutual funds, the government started addressing a proverb, *"Mutual Fund investments are subject to market risks, read all scheme-related documents carefully"*. The NAVs of the schemes may go up or down depending upon the factors and forces affecting the securities market including the fluctuations in the interest rates. Simply it means that Mutual funds, like all securities, are vulnerable to market, or systemic, risk. This is because no one can anticipate what will happen in the future or whether a particular asset will appreciate or depreciate. No investment is risk-free since the market cannot be correctly predicted or entirely controlled.

Each AMC has its method of reporting and representing data, which has shown to be a significant disadvantage when it comes to evaluating and interpreting the data. Also, there is the issue of fund age, with the funds of these individual AMCs being of varying ages. As previously said, only a few funds have reached a maturity of ten years, while only a few have reached a maturity of five years, and only a few have reached the age of three years.

The contribution made by the industry is that the average assets under management (AAUM) in the Indian mutual fund industry were 38.89 lakh crores (INR 38.89 trillion). For the month of January 2022, the Indian mutual fund industry's average assets under management (AAUM) totalled 38,88,571 crores. The AUM of the Indian mutual fund industry has increased by more than 512 times in ten years, from 6.59 trillion on January 31, 2012, to 38.01 trillion on January 31, 2022. The AUM of the mutual fund industry has increased from 17.37 trillion on January 31, 2017, to 38.01 trillion on January 31, 2022, a more than two-fold growth in just five years.

In May 2014, the industry's AUM passed the 10 trillion (ten lakh crore) mark for the first time, and in less than three years, the AUM had expanded more than twofold, passing the 20 trillion (twenty lakh crore) mark for the first time in August 2017. In November 2020, AUM surpassed 30 trillion (30,000 crores) for the first time. As of January 31, 2022, the Industry AUM was 38.01 trillion rupees (38.01 lakh crores). During the month of May 2021, the mutual fund sector reached a milestone of 10 crore folios. As of January 31, 2022, there were 12.31 crore (123.1 million) accounts (or folios in mutual fund lingo), with roughly 9.95 crore folios under Equity, Hybrid, and Solution Oriented Schemes, where the greatest investment is from the retail segment (99.5 million).

From an economic point of view, since 2003, mutual funds have exploded in popularity. Indians typically save up to 30% of their salaried income, which is an extremely significant figure. Mutual funds have long been a popular way for paid people to invest their money. Due to the diversification of mutual fund schemes, more investors have been able to join and pool their assets. Mutual funds provide diversity or access to a broader range of investments than a single investor could afford. When an investor invests in a group, the investor can benefit from economies of scale. The investor's holdings grow because of monthly donations because they are less volatile, and funds are more liquid.

Mutual funds are categorized in various categories, e.g., Large Cap Funds, Multi Cap Funds, Mid Cap Funds, Small Cap Funds, ELSS, Index Funds, Balanced Funds, Debt Funds, and Liquid Funds, etc. This study is about Small Cap and Large Cap Mutual Funds, Banking & Financial Sector funds. All these fund categories are very popular among investors. However, an investor's orientation towards risk and return along with the duration of investment plays a dominant factor to decide which cap to invest in. Generally, Investors with shorter time horizons, lower threshold towards risk and consideration for steady returns prefer to invest in Large Cap Funds.

This research chapter analyses the performances of Small Cap, Large Cap and Banking & Financial Sector funds of Nippon India AMC, ICICI Prudential AMC and Tata AMC aiming that how a Large Cap AMC, Mid Cap AMC and a Short Cap AMC generate good returns from its investment for their investors.

15.2 Objectives of the Study

The objective of the study is to analyze the performances of Mutual Funds offered by different AMCs [Large cap (Nippon India AMC), Medium cap (ICICI Prudential AMC) and small cap AMC (Tata AMC)] in the field of Mutual Fund industry.

15.2.1 Significance of the Study

The analysis highlights the relevance of Nippon India AMC, ICICI Prudential AMC, and Tata AMC's Small Cap, Large Cap, and Banking & Financial Sector funds. The investors' faith in these funds, into which they put their savings from earned income. The analysis also illustrates the performance of the individual AMCs, with Nippon India being classified as a Large-cap AMC, ICICI Prudential AMC as a Mid-cap AMC, and Tata AMC as a Small-cap AMC. Their AUM reflects the interest of investors, which is followed by performance and investment plans established by professional fund managers.

15.3 Literature Review

To undertake a thorough analysis on the subject, the following research papers discuss the mutual funds business and are written by various well-known authors.

Table 1: Review of literature.

Kandi et al. (2022)	The study evaluates the performance of small-cap, mid-cap mutual funds in India. The study was analysed using a sample of 20 funds with ten in each group, during a nine-year period from 2013–2014 to 2021–2022. The study concluded that, among other financial instruments, mutual funds offer the lowest risk and highest return to investors.
Mathur (2021)	The study aims to compare the performance of well-known multi-cap and large-cap funds based on their returns. For this reason, the performance of ten well-known funds in both categories was examined during a five-year period. Their performance was also compared to two of India's most diversified benchmark indices, the BSE 200 and the Nifty 500. The researchers also wanted to see if there was a substantial variation in the performance of these funds.

Table 1 contd. ...

...Table 1 contd.

Tripathi and Japee (2020)	This study focuses on the performance of selected equities (large-cap, mid-cap, small-cap) open-end fund schemes in terms of a risk-return relationship, as offered by various fund companies in India. The main goal of this study is to examine the financial performance of a few open-end fund schemes using statistical measures such as Jenson's alpha, beta, standard deviation, and Sharpe ratio. In a highly volatile market, the researcher found that 10 out of 15 funds fared well.
Tripathi and Japee (2020)	This study focuses on the performance of selected equities (large-cap, mid-cap, small-cap) open-end fund schemes in terms of a risk-return relationship, as offered by various fund companies in India. The main goal of this study is to examine the financial performance of a few open-end fund schemes using statistical measures such as Jenson's alpha, beta, standard deviation, and Sharpe ratio.
Keim (2018)	Investors seeking low-cost exposure to certain parts of the equity (or bond) market are increasingly turning to indexed and passive equity mutual funds, which are designed to deliver the returns and risk of a benchmark index, such as the S&P 500 or the Russell 2000 Index. Duplication is the classic way of creating an index fund.
Agarwal and Mirza (2018)	The research included measuring the performance of selected mutual schemes based on risk and return and comparing the performance of the selected schemes with benchmark index to see whether the scheme is outperforming or underperforming the benchmark. In addition, funds were ranked based on the performance and strategies were suggested to invest in a mutual fund.
Pandow (2017)	The study focusses on the industry drawbacks like low penetration ratio, lack of product differentiation, lack of investor awareness and ability to communicate value to customers, lack of interest of retail investors towards mutual funds and evolving nature of the industry. Based on the analysis, by using Treynor ratio, the study suggests that if the industry utilizes its potential fully, it has to address these challenges.
Rangasamy and Sathiya (2017)	The main objective of the study was to analyze the risk and return of the schemes and to evaluate the performance of equity, debt and balanced schemes of selected mutual funds using Treynor, Sharpe, Jensen measure, etc., containing in the portfolio. The study shows the evaluation of various mutual fund schemes with respect to four financial years (2012–2016).
Cao et al. (2017)	The research adds to the body of knowledge on mutual fund styles, drift, as well as the value and relevance of the investment objectives of mutual funds. First, they showed that there is a perplexing empirical regularity. The large-cap allocation of small-cap funds varies by 63% points across ten deciles. Second, they showed that big deviations from fund objectives do not result in greater out-of-sample performance throughout the course of our sample period. In the years after the technology bubble, however, we show that deviations from fund objectives lead to better performance.

Table 1 contd. ...

...Table 1 contd.

Choudhary and Chawla (2014)	The paper conducted research on the topic performance evaluation of mutual funds: a study of selected diversified equity mutual funds in India. Through this study an attempt has been made to analyze the performance of growth-oriented equity diversified schemes based on return and risk evaluation. The analysis was achieved by assessing various financial tests like Average Return, Sharpe Ratio, Treynor Ratio, Standard Deviation, Beta and Coefficient of Determination.
Cao et al. (2017)	The study emphasizes the significance of conducting thorough research when selecting a fund, as well as the need for improved financial regulation that matches investor and fund manager interests.
Nandhini and Rathnamani (2017)	The study focuses on the performance of selected equity large and small cap mutual fund schemes and they were analyzed with various risk return measurement tools such as alpha, beta, standard deviation and Sharpe ratio.
Otten and Reijnders (2016)	The performance of mutual funds that specialize in investing in smaller companies in the United Kingdom is examined in this article. In comparison to mutual funds that invest in major business stock, research on the small cap fund area of the market is scarce and plagued by methodological flaws.
Damayanti and Cintyawati (2016)	The study examined that there are several factors that are considered to affect the performance of mutual funds such as asset under management (AUM), fund age, past performance, asset allocation, turn of the year effect, equity funds with blue chip or non-blue-chip stocks, equity funds owned by insurance or non-insurance companies, external factors such as the rupee against the US dollar (exchange rate) and investors behavior, etc.
Kumar and Kumar (2016)	The study examined the research with the prominent objective to determine the appropriate Benchmark Index that consists of appropriate asset classes of securities to pave the way for precise estimation. The study considers Tax Planning (Equity Linked Savings Scheme-ELSS) funds and selected indices of National Stock Exchange and Bombay Stock Exchange. The study reveals that broad based indices that consist of Large cap, Mid cap, and Small cap asset classes would be an appropriate benchmark for evaluating the performance of ELSS funds.
Otten and Reijnders (2016)	The study examined the long-term consistency in small cap fund performance and whether the findings of large cap fund performance studies can be applied to the small cap segment the study used a condition multi factor model fitted to the small firm stock universe to solve earlier methodological shortcomings.
Narayanasamy and Rathnamani (2015)	The main objective of the research work was to create an analysis of the financial performance of selected mutual fund schemes through the statistical parameters such as alpha, beta, standard deviation, r-squared, Sharpe ratio, etc.
Rai et al. (2014)	The purpose of this study is to examine the performance of large cap, mid cap & small-cap equity mutual funds in India, as well as their benchmark indices, over one, three, and five years. In total, 40 equities open ended mutual fund schemes are chosen—22 large-cap and 18'mid & small-cap' schemes—and their annualized returns are compared over the last 1, 3, and 5 years.

Table 1 contd. ...

...Table 1 contd.

Ashraf and Sharma (2014)	A study made by, Husain and Sharma (2014), over a five-year period compared the performance of equities mutual funds to the risk-free rate and benchmark return. According to the risk return study, three of the ten schemes underperformed the market, while seven had lower total risk than the market and all had greater returns than risk free rates. At a 5% level of significance, regression analysis revealed that benchmark market index return has a statistically significant impact on mutual fund return.
Philips and Kinniry (2010)	A study by Philips and Kinniry (2010) looked at mutual fund evaluations and future performance. Their paper tackles two issues related to mutual fund rating systems. First, why do index funds tend to have an average rating based on comparable quantitative metrics? Second, is a given performance rating actionable? The article specifically looked at whether higher-rated funds could beat lower-rated funds in the future.
Nitzsche et al. (2006)	A study has been conducted to study the performance of funds with respect to benchmarks. The study shows the mutual fund performance in the United States and the United Kingdom, where it was discovered that 2–5% of funds truly beat the benchmarks, whereas 20–40 % of funds underperform the benchmarks.
Friend et al. (1962)	A study has been conducted to study the returns of funds with respect to benchmarks Friend et al. (1972) discovered that mutual fund returns were almost identical to those of the benchmark index in one of the first studies on mutual fund performance. It was observed that because managed funds were unable to beat the benchmark index, market efficiency existed in the stock markets.
Jensen (1968)	A study has been conducted to study the performance of open-ended mutual funds with respect to benchmarks. Jensen (1968) looked at the performance of 115 open-end mutual funds from 1945 to 1964 and discovered that the funds' returns were on average no better than the benchmark (S&P 500) index.
Sharpe (1966)	A study has been conducted to study the performance and return of funds with respect to benchmarks. Sharpe (1966) used conducted a study to variability ratio, 'the Sharpe ratio', to compare the performance of 34 mutual funds with the Dow-Jones industrial average from 1954 to 1963. Observations came up that the fund's total performance was worse to the Dow-Jones index.

15.4 Research Gap

The empirical review in these chapters clearly indicates that some research gaps have been identified while analyzing the papers—Performance and analysis of Mid Cap funds, Sectoral/Thematic funds were not analyzed, Research from Asian market perspective was not analyzed properly, research on Benchmark Index like S&P 500, Dow Jones was not included in-depth, several parameters like Jensen's ratio were not included. In addition to them, to compare the Large Cap funds, a Comparison of 10-year performance was a gap along with the Ask rate and Bid rate of foreign exchanges and the INR.

15.5 Research Methodology

For this study, quantitative data has been collected. Data that can be counted or measured in numerical values is referred to as quantitative data. Apart from this, many factors were considered when conducting the study and collecting the data, including the type of data, the universe in which it originates, the nature of the data source, sample procedures, critical variables, and so on.

15.5.1 Data Source

To undertake a thorough and comprehensive analysis on the topic, data has been collected of Small Cap (Equity), Large Cap (Equity) and Banking & Financial Sector funds (Sectoral/Thematic) of Nippon India AMC, ICICI Prudential AMC and Tata AMC. Their performance in the market and returns are given to inventors would be analyzed keeping S&P 500 as the benchmark.

The study covers the period of 10 years from 2011 to 2021. The study is wholly designed on secondary data and has been obtained from the official websites of the respective AMCs, i.e., Nippon India AMC, ICICI Prudential AMC and Tata AMC. The data has been collected from their factsheets and fund performance reports. Apart from this, data have also been collected from Moneycontrol, Groww, The Association of Mutual Funds in India (AMFI), Securities & Exchange Board of India (SEBI) and The Ministries of Statistics and Programme Implementation (MoSPI).

15.5.2 Data Analysis Techniques

This study is a combination of the nature of fund and fund performance. To analyze the performance and yield of funds given to investors, the following statistical tests have been kept in mind:

1. **Time Series Analysis**: A time series analysis is a method of studying a collection of data points over a period. Instead of capturing data points intermittently or arbitrarily, time series analyzers record data points at constant intervals over a predetermined length of time.

2. **Standard Deviation (SD):** A standard deviation (or σ) is a measure of data dispersion in proportion to the mean. Data are grouped around

the mean when the standard deviation is low, while data are more spread out when the standard deviation is high.

3. **T-test:** A t-test is an inferential statistic that is used to see if there is a significant difference in the means of two groups that are related in some way.

15.6 Data Analysis

Table 2: Nippon India Small Cap Fund.

	Nippon- Direct Plan	Nippon-Regular Plan
Mean	2213.48	2287.14
Variance	5144420.131	6596362.954
Observations	10	10
Pearson Correlation	0.998675145	
Hypothesized Mean Difference	1	
Df	9	
t Stat	−0.72667262	
P(T<=t) one-tail	0.242948036	
t Critical one-tail	1.833112933	
P(T<=t) two-tail	0.485896073	
t Critical two-tail	2.262157163	

Source: Own Compilation

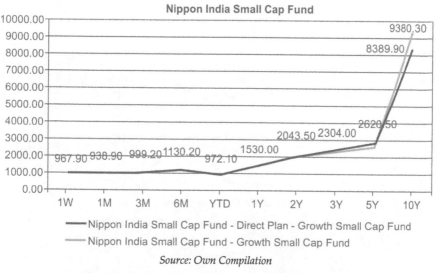

Source: Own Compilation

Figure 1: Performance of Nippon India Small Cap Fund.

As the Pearson Correlation is 0.99 (closer to 1), it indicates a negative relationship between the two variables. While both direct and regular funds in this plan show significant growth in their investments, the difference arises when brokerage fees are deducted in the case of regular investments, resulting in a sum of Rs. 9380.30 in the case of a direct plan and Rs. 8389.90 in the case of the regular plan both invested for 10 years each (Principal investment amount has been taken as Rs. 1000).

Table 3: Nippon India Large Cap Fund.

	Nippon- Direct Plan	Nippon-Regular Plan
Mean	1257.888889	1238.755556
Variance	128643.2761	108553.2003
Observations	9	9
Pearson Correlation	0.99979175	
Hypothesized Mean Difference	0.05	
df	8	
t Stat	1.906697453	
P(T<=t) one-tail	0.046501666	
t Critical one-tail	1.859548038	
P(T<=t) two-tail	0.093003331	
t Critical two-tail	2.306004135	

Source: Own Compilation

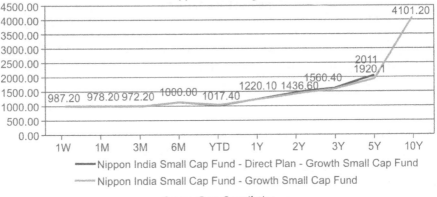

Source: Own Compilation

Figure 2: Performance of Nippon India Small Cap Fund.

As the Pearson Correlation is 0.99 (closer to 1), it indicates a negative relationship between the two variables. While both direct and regular funds in this plan show significant growth in their investments, the difference arises when brokerage fees are deducted in the case of regular investments, resulting in a sum of Rs. 1920.10 in the case of the direct plan for 5 years and Rs. 4101.20 in the case of a regular plan for 10 years. (Principal investment amount has been taken as Rs. 1000).

Table 4: Nippon India Banking and Financial services Fund.

	Nippon- Direct Plan	Nippon-Regular Plan
Mean	1416.87	1462.82
Variance	453714.8801	717605.9329
Observations	10	10
Pearson Correlation	0.993735119	
Hypothesized Mean Difference	0.05	
df	9	
t Stat	−0.753560845	
P(T< = t) one-tail	0.235183226	
t Critical one-tail	1.833112933	
P(T< = t) two-tail	0.470366453	
t Critical two-tail	2.262157163	

Source: Own Compilation

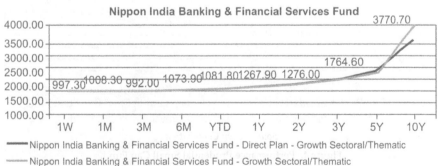

Nippon India Banking & Financial Services Fund

Source: Own Compilation

Figure 3: Performance of Nippon Banking and Financial Services Fund.

As the Pearson Correlation is 0.99 (closer to 1), it indicates a negative relationship between the two variables. While both direct and regular funds in this plan show significant growth in their investments, the difference arises when brokerage fees are deducted in the case of regular investments, resulting in a sum of Rs. 3770.70 in the case of a direct plan and Rs. 3178.60 in the case of the regular plan both invested for 10 years each. (Principal investment amount has been taken as Rs. 1000).

Table 5: ICICI Prudential Small Cap Fund.

	ICICI Pru-Direct Plan	ICICI Pru-Regular Plan
Mean	1442.777778	1408.122222
Variance	363561.8494	310231.3369
Observations	9	9
Pearson Correlation	0.999807165	
Hypothesized Mean Difference	0.05	
df	8	
t Stat	2.191871793	
P(T<=t) one-tail	0.029873184	
t Critical one-tail	1.859548038	
P(T<=t) two-tail	0.059746369	
t Critical two-tail	2.306004135	

Source: Own Compilation

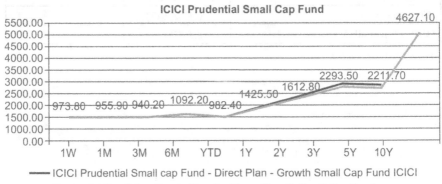

Source: Own Compilation

Figure 4: Performance Prudential Small Cap Fund.

As the Pearson Correlation is 0.99 (closer to 1), it indicates a negative relationship between the two variables. While both direct and regular funds in this plan show significant growth in their investments, the difference arises when brokerage fees are deducted in the case of regular investments, resulting in a sum of Rs. 2211.70 in the case of the direct plan for 5 years and Rs. 4627.10 in the case of a regular plan for 10 years. (Principal investment amount has been taken as Rs. 1000).

Table 6: ICICI Prudential Large Cap Fund.

	ICICI Pru- Direct Plan	ICICI Pru-Regular Plan
Mean	1283.833333	1268.277778
Variance	153127.925	134912.1844
Observations	9	9
Pearson Correlation	0.999800874	
Hypothesized Mean Difference	0.05	
df	8	
t Stat	1.847700247	
P(T<=t) one-tail	0.050917698	
t Critical one-tail	1.859548038	
P(T<=t) two-tail	0.101835395	
t Critical two-tail	2.306004135	

Source: Own Compilation

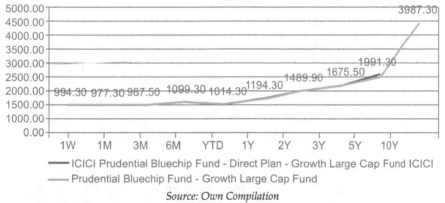

Source: Own Compilation

Figure 5: Performance of ICICI Prudential Large Cap Fund.

As the Pearson Correlation is 0.99 (closer to 1), it indicates a negative relationship between the two variables. While both direct and regular funds in this plan show significant growth in their investments, the difference arises when brokerage fees are deducted in the case of regular investments, resulting in a sum of Rs. 1991.30.10 in the case of the direct plan for 5 years and Rs. 3987.30 in the case of a regular plan for 10 years. (Principal investment amount has been taken as Rs. 1000).

Table 7: ICICI Prudential Banking and Financial services Fund.

	ICICI Pru-Direct Plan	ICICI Pru-Regular Plan
Mean	1195.644444	1176.711111
Variance	88764.15278	71952.40361
Observations	9	9
Pearson Correlation	0.999764787	
Hypothesized Mean Difference	0.05	
Df	8	
t Stat	1.868395971	
P(T<=t) one-tail	0.049324945	
t Critical one-tail	1.859548038	
P(T<=t) two-tail	0.098649891	
t Critical two-tail	2.306004135	

Source: Own Compilation

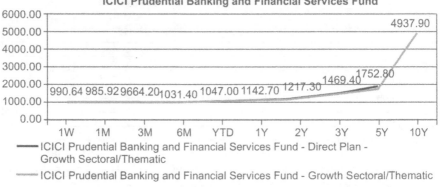

Source: Own Compilation

Figure 6: Performance of ICICI Prudential Banking and Financial Services Fund.

As the Pearson Correlation is 0.99 (closer to 1), it indicates a negative relationship between the two variables. While both direct and regular funds in this plan show significant growth in their investments, the difference arises when brokerage fees are deducted in the case of regular investments, resulting in a sum of Rs. 1752.80 in the case of direct plan for 5 years and Rs. 4937.90 in the case of regular plan for 10 years (Principal investment amount has been taken as Rs. 1000).

Table 8: TATA Small Cap Fund .

	Tata-Direct Plan	Tata-Regular Plan
Mean	1338.3125	1308.25
Variance	290146.7241	245188.6714
Observations	8	8
Pearson Correlation	0.999772898	
Hypothesized Mean Difference	0.05	
Df	7	
t Stat	1.892347201	
P(T<=t) one-tail	0.050164295	
t Critical one-tail	1.894578605	
P(T<=t) two-tail	0.10032859	
t Critical two-tail	2.364624252	

Source: Own Compilation

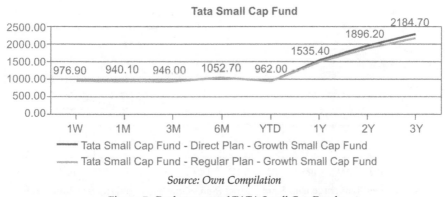

Source: Own Compilation

Figure 7: Performance of TATA Small Cap Fund.

As the Pearson Correlation is 0.99 (closer to 1), it indicates a negative relationship between the two variables. While both direct and regular funds in this plan show significant growth in their investments, the difference arises when brokerage fees are deducted in the case of regular investments, resulting in a sum of Rs. 2309.40 in the case of a direct plan and Rs. 2184.70 in the case of the regular plan both invested for 3 years each (Principal investment amount has been taken as Rs. 1000).

Table 9: TATA Large Cap Fund.

	Tata-Direct Plan	Tata-Regular Plan
Mean	1252.166667	1231.211111
Variance	131864.0525	109342.0336
Observations	9	9
Pearson Correlation	0.999383164	
Hypothesized Mean Difference	0.05	
Df	8	
t Stat	1.809038655	
P(T<=t) one-tail	0.054024699	
t Critical one-tail	1.859548038	
P(T<=t) two-tail	0.108049397	
t Critical two-tail	2.306004135	

Source: Own Compilation

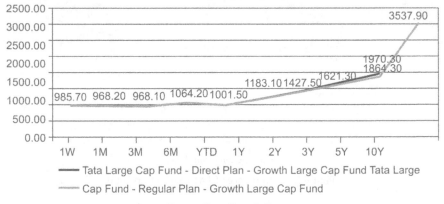

Source: Own Compilation

Figure 8: Performance of TATA Large Cap Fund.

As the Pearson Correlation is 0.99 (closer to 1), it indicates a negative relationship between the two variables. While both direct and regular funds in this plan show significant growth in their investments, the difference arises when brokerage fees are deducted in the case of regular investments, resulting in a sum of Rs. 1970.30 in the case of the direct plan for 5 years and Rs. 3537.90 in the case of a regular plan for 10 years (Principal investment amount has been taken as Rs. 1000).

Table 10: TATA Banking and Financial Services Fund.

	Tata-Direct Plan	Tata-Regular Plan
Mean	1208.022222	1174.833333
Variance	138101.7844	102469.36
Observations	9	9
Pearson Correlation	0.999771972	
Hypothesized Mean Difference	0.05	
Df	8	
t Stat	1.910534244	
P(T<=t) one-tail	0.046227517	
t Critical one-tail	1.859548038	
P(T<=t) two-tail	0.092455035	
t Critical two-tail	2.306004135	

Source: Own Compilation

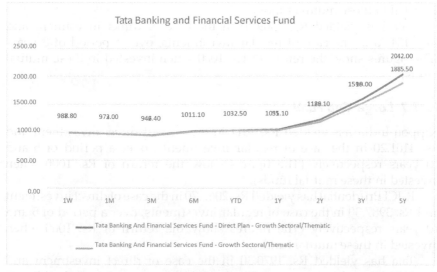

Source: Own Compilation

Figure 9: Performance of TATA Banking and Financial Services Fund.

As the Pearson Correlation is 0.99 (closer to 1), it indicates a negative relationship between the two variables. While both direct and regular funds in this plan show significant growth in their investments, the difference arises when brokerage fees are deducted in the case of regular investments, resulting in a sum of Rs. 2042.00 in the case of the direct plan and Rs. 1885.5 in the case of the regular plan both invested for 5 years each (Principal investment amount has been taken as Rs. 1000).

Thus, it is observed that the Large Cap Company, i.e., Nippon India AMC, gave the best returns to its investors in both direct and regular types of funds followed by Mid Cap Company, i.e., ICICI Prudential AMC and Small Cap Fund, i.e., Tata AMC.

15.7 Findings

15.7.1 Small Cap fund

Nippon India has yielded Rs. 9380.30 in the case of direct investment and Rs. 8389.90 in the case of regular investments, over a period of 10 years (The figures show the return of Rs. 1000 when invested in these mutual funds).

ICICI Prudential has yielded Rs. 2336.50 in the case of direct investment and Rs. 4627.10 in the case of regular investments, over a period of 5 and 10 years respectively (The figures show the return of Rs. 1000 when invested in these mutual funds).

Tata has yielded Rs. 2309.40 in the case of direct investment and Rs. 2184.70 in the case of regular investments, over a period of 3 years (The figures show the return of Rs. 1000 when invested in these mutual funds).

15.7.2 Large Cap Fund

Nippon India has yielded Rs. 2011 in the case of direct investment and Rs. 4101.20 in the case of regular investments, over a period of 5 and 10 years respectively (The figures show the return of Rs. 1000 when invested in these mutual funds).

ICICI Prudential has yielded Rs. 2067.70 in the case of direct investment and Rs. 3987.30 in the case of regular investments, over a period of 5 and 10 years respectively (The figures show the return of Rs. 1000 when invested in these mutual funds).

Tata has yielded Rs. 1970.30 in the case of direct investment and Rs. 3537.30 in the case of regular investments, over a period of 5 and 10 years respectively (The figures show the return of Rs. 1000 when invested in these mutual funds).

15.7.3 Banking and Financial Services Fund

Nippon India has yielded Rs. 3178.60 in the case of direct investment and Rs. 3770.70 in the case of regular investments, over a period of 10 years

(The figures show the return of Rs. 1000 when invested in these mutual funds).

ICICI Prudential has yielded Rs. 1844.90 in the case of direct investment and Rs. 4737.90 in the case of regular investments, over a period of 5 and 10 years respectively (The figures show the return of Rs. 1000 when invested in these mutual funds).

Tata has yielded Rs. 2042 in the case of direct investment and Rs. 1885.50 in the case of regular investments, over a period of 5 years (The figures show the return of Rs. 1000 when invested in these mutual funds).

15.8 Conclusion

A mutual fund is a powerful investment choice with the potential to provide investors with long-term riches. Mutual funds include plans for all kinds of life goals, from building a wealth pool to retirement. Diversification, low cost, flexibility to invest in lesser quantities, and expert fund administration are all advantages of this approach. When combined with an online investment platform, an investor has a fantastic tool that makes mutual fund investing simple and painless.

As observed in the data, the Large Cap Company, i.e., Nippon India AMC, gave the best returns to its investors in both direct and regular types of funds followed by Mid Cap Company, i.e., ICICI Prudential AMC and Small Cap Fund, i.e., Tata AMC.

According to the research, mutual fund schemes in all three categories, i.e., Small Cap Funds, Large Cap Funds and Banking & Financial Services Fund have delivered positive returns over time while posing a fair risk. As a result, it is extremely safe to infer that they are an excellent investment alternative. The average monthly returns generated by the funds in each category are quantitatively different in terms of performance. There is no statistically significant difference in the means of average monthly returns of funds in all these categories. In terms of returns, there is no substantial difference between these funds and the NIFTY 500, nor between these funds and the BSE 200.

Thus, it implies that the bigger the AuM (Fund Size) of an AMC's fund, the higher the trust of investors which leads to better performance of the funds as the strategies of investment into different sectors are being prepared by experienced fund managers. The fund management diversifies the investment by spreading it among equities of firms from diverse industries and sectors. When one asset class underperforms, the other sectors can compensate to protect investors from losses. They have also been shown to outperform inflation because the amount invested is compounded annually using the CAGR (Compound Annual Growth Rate) approach. They also assist an investor in saving money on taxes.

An investor can invest in ELSS mutual funds, which are tax-saving mutual funds that qualify for a tax deduction of up to Rs 1.5 lakh per year under Section 80C of the Income Tax Act, 1961. Even though Long-Term Capital Gains (LTCG) over Rs. 1 lakh are subject to a 10% tax, they have consistently outperformed other tax-saving products in recent years.

15.9 Limitations of the Study

There were certain limitations to the research. To begin with, data collection was a challenge. Each AMC has its method of reporting and representing data, which has shown to be a significant disadvantage when it comes to evaluating and interpreting the data.

Second, there was the issue of fund age, with the funds of these individual AMCs being of varying ages. As previously said, only a few funds have reached a maturity of ten years, while only a few have reached a maturity of five years, and only a few have reached the age of three years. The age of direct and regular funds also varies, with only a few reaching maturities of 10 years and others only 5 years.

Third, there were few papers available on this topic, making it difficult to conduct this research and make it a quality study.

And IT infrastructure of the AMCs is not adequate as each AMC has its way of presenting and delivering the data to the investors to attract them.

15.10 Managerial Implications

After analyzing the research study, it was found that many players are befitted from the mutual fund industry. Mainly those players are the investors who invest their money in the funds along with the brokers who get a commission on each transaction as requested by the investor. Along with them, it also helps the mutual fund industry to grow as the official website of mutual funds, i.e. The Association of Mutual Funds in India (AMFI), published that, the contribution made by the industry is that the average assets under management (AAUM) in the Indian mutual fund industry were 38.89 lakh crores (INR 38.89 trillion). For the month of January 2022, the Indian mutual fund industry's average assets under management (AAUM) totalled 38,88,571 crores. The AUM of the Indian mutual fund industry has increased by more than 512 times in ten years, from 6.59 trillion on January 31, 2012, to 38.01 trillion on January 31, 2022. The AUM of the mutual fund industry has increased from 17.37 trillion on January 31, 2017, to 38.01 trillion on January 31, 2022, a more than two-fold growth in just five years.

In May 2014, the industry's AUM passed the 10 trillion (ten lakh crore) mark for the first time, and in less than three years, the AUM had expanded more than twofold, passing the 20 trillion (twenty lakh crore) mark for the first time in August 2017. In November 2020, AUM surpassed 30 trillion (30,000 crores) for the first time. As of January 31, 2022, the Industry AUM was 38.01 trillion rupees (38.01 lakh crores). During the month of May 2021, the mutual fund sector reached a milestone of 10 crore folios. As of January 31, 2022, there were 12.31 crore (123.1 million) accounts (or folios in mutual fund lingo), with roughly 9.95 crore folios under Equity, Hybrid, and Solution Oriented Schemes, where the greatest investment is from the retail segment (99.5 million). This also affects the Indian economy as well, in a positive manner and will continue to rise more in the coming years.

References

Agarwal, S., and N. Mirza. 2018. A study on the risk-adjusted performance of mutual funds industry in India. *Review of Innovation and Competitiveness* 3(1): 75–94.

Ashraf, S. H., and D. Sharma. 2014. Performance evaluation of Indian equity mutual funds against established benchmarks index. *International Journal of Accounting Research* 2.

Ayaluru, M. P. 2016. Performance analysis of mutual funds: Selected reliance mutual fund schemes. *Parikalpana: KIIT Journal of Management* 12(1): 52–62.

Bengtsson, E. 2013. Shadow banking and financial stability: European money market funds in the global financial crisis. *Journal of International Money and Finance* 32: 579–594.

Cao, C., P. Iliev, and R. Velthuis. 2017. Style drift: Evidence from small-cap mutual funds. *Journal of Banking & Finance* 78: 42–57.

Choudhary, V., and P. S. Chawla. 2014. Performance evaluation of mutual funds: A study of selected diversified equity mutual funds in India. *International Conference on Business, Law and Corporate Social Responsibility (ICBLCSR'14)* Oct 1–2, 2014 Phuket (Thailand), 82–85.

Damayanti, S. M., and C. Cintyawati. 2016. Developing an integrated model of equity mutual funds performance: Evidence from the Indonesian mutual funds market. *GSTF Journal on Business Review (GBR)* 4(1): 124–135.

Friend, I., F. E. Brown, E. S. Herman, and D. Vickers. 1962. A study of Mutual Funds. US Government Printing Office, Washington, DC.

Gallo, J. G., V. P. Apilado, and J. W. Kolari. 1996. Commercial bank mutual fund activities: Implications for bank risk and profitability. *Journal of Banking & Finance* 20(10): 1775–1791.

Gorman, L. 2003. Conditional performance, portfolio rebalancing, and momentum of small-cap mutual funds. *Review of Financial Economics* 12(3): 287–300.

Haslem, J. A., and C. A. Scheraga. 2006. Data envelopment analysis of morningstar's small-cap mutual funds. *The journal of Investing* 15(1): 87–92.

Jensen, M. C. 1968. The performance of mutual funds in the period 1945–1964. *The Journal of Finance* 23: 389–416.

Kandi, V. S., K. Harshitha, and A. Dhyan. 2022. Performance of mutual funds in india: comparative analysis of small-cap and mid-cap mutual funds. *Academy of Marketing Journal* 26(4).

Khan, M. M., and M. I. Bhatti. 2008. Development in Islamic banking: a financial risk-allocation approach. *The Journal of Risk Finance*.

Keim, D. B. 1999. An analysis of mutual fund design: the case of investing in small-cap stocks. *Journal of Financial Economics* 51(2): 173–194.

Kumar, V., and A. Kumar. 2016. Construction of appropriate benchmark index for mutual funds: Specific reference to tax saver funds. *International Journal of Financial Management* 2(1): 74–90.

Mathur, P. 2021. Comparative Analysis of performance of mutual funds: a study of prominent multi cap and large cap funds. *Contemporary Issues and Recent Advances in Management, Commerce, Economics*, 23.

Nandini, R., and V. Rathnamani. 2017. A study on the performance of equity mutual funds (with special reference to equity large cap and mid cap mutual funds). *IOSR Journal of Business and Management*, 19(2): 67–72.

Narayanasamy, R. V., and Rathnamani. 2015. Performance evaluation of equity mutual funds (On Selected Equity Large Cap Funds). *International Journal of Business and Management Invention* 2(4): 18–24.

Nitzsche, D., K. Cuthbertson, and N. O'Sullivan. 2006. Mutual fund performance. Retrieved from https://www.evidenceinvestor.com/wp-content/uploads/2016/08/Mutual-Fund-Performance-Nitzsche-Dirk-Cuthbertson-Keith-and-OSullivan-Niall-2006.pdf.

Otten, R., and M. Reijnders. 2012. *The Performance of Small-cap Mutual Funds: Evidence for the UK.* Working paper Maastricht University Department of Finance.

Otten, R., and M. Reijnders. 2016. *The Performance of Small Cap Mutual Funds: Evidence for the UK.* Retrieved from https://efmaefm.org/0efmsymposium/2012/papers/Otten.pdf.

Pandow, B. 2017. Performance of Mutual Funds in India. Retrieved from file:///C:/Users/307084/Downloads/SSRN-id2925049.pdf.

Philips, C. B., and F. M. Kinniry. 2010. Mutual fund ratings and future performance. *Economics*.

Rai, R. S., T. V. Raman, and G. Shreekant. 2014. Comparing returns between large and mid & small cap equity mutual funds in India. *International Journal of Applied Research* 4(12): 324–328.

Rangasamy, S., and P. Sathiya. 2017. Trend and performance of selected mutual funds. *International Research Journal of Engineering and Technology* 4(2): 1651–1654.

Sharpe, W. F. 1966. Mutual fund performance. *The Journal of Business* 39: 119–138.

Tripathi, S., and D. G. P. Japee. 2020. Performance evaluation of selected equity mutual funds in India. *GAP GYAN-A Global Journal of Social Science*.

Index